中国植物病理学会第十二届青年学术研讨会论文选编

植物病理学研究进展

吴学宏　主编

U0213132

中国农业大学出版社
·北京·

内容简介

本书共收集了135篇研究论文、简报、摘要和综述。涉及植物病原真菌及真菌病害、病毒及病毒病害、原核生物及其病害、线虫等其他病原及其病害、生物防治、病害流行及预测、种子病理与杀菌剂、抗病性及抗病育种、病害综合防治、分子生物学及其应用、信息技术及其应用等方面，基本上反映了近两年来我国植物病理学青年工作者在植物病理学各分支学科基础理论、应用基础和植物病害防治实践等方面所取得的研究进展。

本书可供植物病理学、农学、生物学等相关专业师生、科研人员参考。

图书在版编目(CIP)数据

植物病理学研究进展/吴学宏主编.—北京：中国农业大学出版社，2015.10
ISBN 978-7-5655-1421-0

Ⅰ.①植… Ⅱ.①吴… Ⅲ.①植物病理学－文集 Ⅳ.①S432.1－53

中国版本图书馆CIP数据核字(2015)第238384号

书　名　植物病理学研究进展
作　者　吴学宏　主编

策划编辑　赵　中　　**责任编辑**　韩元凤
封面设计　郑　川
出版发行　中国农业大学出版社
社　址　北京市海淀区圆明园西路2号　　**邮政编码**　100193
电　话　发行部 010-62818525,8625　　读者服务部 010-62732336
编辑部 010-62732617,2618　　出　版　部 010-62733440
网　址　http://www.cau.edu.cn/caup　　**E-mail** cbsszs @ cau.edu.cn
经　销　新华书店
印　刷　涿州市星河印刷有限公司
版　次　2015年10月第1版　2015年10月第1次印刷
规　格　787×1092　16开本　11.25印张　278千字
定　价　60.00元

图书如有质量问题本社发行部负责调换

中国植物病理学会第十二届青年学术研讨会
组织委员会

顾　问　彭友良

主　任　吴学宏

副主任　王海光　燕继晔　杨　俊　李向东

委　员　（按姓氏笔画为序）

王海光　王晓杰　田延平　刘文德　刘永锋　李向东　李　燕
陈旭君　吴学宏　杨　俊　汪　涛　周　涛　原雪峰　曹志艳
梁春浩　谢甲涛　窦道龙　燕继晔　魏太云

编委会

主　编　吴学宏

副主编　王海光　杨　俊　燕继晔

编　委　（按姓氏笔画为序）

王海光　王晓杰　田延平　李向东　吴学宏　杨　俊

原雪峰　曹志艳　梁春浩　窦道龙　燕继晔

前　言

中国植物病理学会第十二届青年学术研讨会论文选编《植物病理学研究进展》共收集了135篇研究论文、简报和摘要。内容涉及植物病原真菌及真菌病害、病毒及病毒病害、原核生物及其病害、线虫等其他病原及其病害、生物防治、病害流行及预测、种子病理与杀菌剂、抗病性及抗病育种、病害综合防治、分子生物学及其应用、信息技术及其应用等方面，基本上反映了近两年来我国植物病理学青年工作者在植物病理学各分支学科基础理论、应用基础和植物病害防治实践等方面所取得的研究进展。

本次大会由中国植物病理学会青年委员会主办，以"青年植病工作者与科技创新"为会议主题。大会的召开和论文集的出版得到了中国科学技术协会和中国植物病理学会的资助。挂靠单位中国农业大学给予了关心和指导。我国植物病理学界多位专家、教授给予了关心、指导和帮助。大会承办单位山东农业大学植物保护学院和山东植物病理学会，以及编辑委员会的同志们付出了辛苦的劳动，中国农业大学出版社给予了大力支持和帮助。在此，我们表示衷心的感谢！

由于大会征集论文时间仓促，同时，本着尊重作者意愿和文责自负的原则，对论文的内容一般未作改动，仅在编辑体例上进行了处理。因此，错误和不足之处在所难免，敬希论文作者和读者批评指正！

编　者

2015年10月

目　录

Type Ⅴ myosin FaMyo2B affects asexual and sexual development, reduces pathogenicity, and cooperating with myosin passenger protein FaSmy1 regulates resistance to the fungicide phenamacril in *Fusarium asiaticum*

Xiumei Liu, Yiqiang Cai, Bin Li, Mingguo Zhou*

(College of Plant Protection, Nanjing Agricultural University, Key Laboratory of Pesticide, Nanjing 210095, China)

Fusarium head blight (FHB) or scab of wheat and other small cereal grains caused by *Fusarium graminearum* sensu lato (teleomorph *Gibberella zeae* (Schwein.) Petch) is a disease that causes severe yield and economic losses worldwide (Bai and Shaner, 2004; Goswami and Kistler, 2004). FHB not only reduce grain yield and quality, but can also contaminate grains with a variety of potent mycotoxins that are a threat to human and animal health (Desjardins et al., 2006; Sutton et al., 1982). Although a number of *Fusarium* spp. can cause FHB, the primary etiological agents of this disease belong to the *Fusarium graminearum* species complex of B-trichothecene toxin producers, which contain at least 11 phylogenetic species, including *F. acaciae-mearnsii*, *F. asiaticum*, *F. austroamericanum*, *F. boothii*, *F. brasilicum*, *F. cortaderiae*, *F. gerlachii*, *F. graminearum*, *F. meridionale*, *F. mesoamericanum*, and *F. vorosii*, (11, 32, 34). Different *Fusarium* spp. may be associated with FHB in different regions of the world because of different cropping systems and climatic conditions (28, 36). In China, FHB was first reported in 1936 and FHB epidemics have since become more severe and frequent in the middle and lower regions of the Yangtze River and in the Heilongjiang province in the northeastern region (Chen et al., 2000).

Zhang et al. (2007) analyzed 299 isolates collected from various epidemic regions of China and found that 231 isolates (77.3%) belonged to *F. asiaticum* and the remaining 68 isolates were *F. graminearum*.

Because natural resistance against FHB pathogens is limited, which has severely hampered progress in breeding for resistance with conventional approaches (Chen et al., 2000; Parry et al., 1995; Windels, 2000), the most efficient strategy for the control of FHB is through the application of fungicides during wheat anthesis. The use of a novel cyanoacrylate fungicide phenamacril reduced both the FHB index and mycotoxin level by 80% (Li et al., 2008; Chen and Zhou, 2009; Zhang et al., 2010; Zheng et al., 2015). In vitro, phenamacril-resistant mutants were obtained easily by ultra-violet (UV) irradiation and

* Corresponding author: Mingguo Zhou, E-mail: mgzhou@njau.edu.cn.

fungicide domestication. Most of the resistant mutants belonged to moderately or highly resistance and exhibited similar biological fitness to the wild-type strains. In our previous studies, we found that mutations in myosin-5 confers resistance to phenamacril in *F. graminearum* (Zheng et al., 2015). In *F. graminearum*, the myosin gene family has three members, including FGSG_08719.1, which encodes myo2 (Song et al., 2013); FGSG_07469.1, which encodes myosin-2B; and FGSG_01410.1, which encodes myosin-5. All three of these myosin proteins have conserved "head" regions. The head or motor domain contains binding sites for ATP and actin. To determine whether the other myosin proteins could regulate the resistance to phenamacril in *F. asiaticum*, we evaluated the functions of myosin-2B and myo2 by gene deletion.

Myosins are molecular motors that catalyze an ATP-dependent interaction with actin filaments and generate unidirectional, chemo-mechanical force. Force generation resides in a~80 kDa motor domain that is highly conserved among all myosins. Based on genomic survey and phylogenetic analyses, 31 myosin classes have been defined (Sebe-Pedros et al., 2014). In particular, Class V myosins are processive molecular motors that transport their cargo toward the plus ends of actin filaments. They are involved in numerous membrane trafficking events (Reck-Peterson et al., 2000; Trybus, 2008). *Saccharomyces cerevisiae* has two class V myosins, the essential Myo2 and the nonessential Myo4. While Myo4 mediates the transport of mRNAs and movement of ER tubules, Myo2 plays a major role in the transport of secretory vesicles and segregation of membrane-bounded organelles including vacuoles, peroxisomes, and organelles of the secretory pathway (Matsui, 2003; Pruyne et al., 2004; Weisman, 2006; Fagarasanu et al., 2010). The Myol gene encodes a class Ⅱ myosin that, depending on the strain background, is either essential or nonessential for viability. However, Myol is important for normal cytokinesis and cell wall maintenance in yeast cells (Nitza et al., 2007). A myosin light chain that associates with Myol and Myo2 heavy chains is encoded by the essential Mlcl gene (Stevens and Davis, 1998; Luo et al., 2004). In *S. cerevisiae*, the kinesin-like myosin passenger-protein Smyl transported by myosin V is part of a negative feedback mechanism that detects cable length and prevents overgrowth (Melissa et al., 2011). And the coiled-coil interactions (CCIs) network reveals that Myol and Myo2 are interacting proteins and regulate Smylp, when overexpressed, can partially compensate for defects in the Myo2 mutant, overcoming lethality and restoring polarized growth at restrictive temperature (Wang et al., 2012; Lillie and Brown, 1992, 1994; Zhang et al., 2009).

In *F. graminearum*, Song et al. (2013) identified a type Ⅱ myosin gene, designated as myo2, and demonstrated that the type Ⅱ myosin myo2 is essential for septation, conidiation and sexual reproduction, and plays a significant role in pathogenesis and mycotoxin production. In this paper, we found type Ⅴ myosin gene FaMyo2B in *F. asiaticum* affects asexual and sexual development, reduces pathogenicity, and cooperating with myosin passenger protein gene FaSmyl regulates resistance to the fungicide phenamacril. Our data suggest that FaMyo2B and Famyo2 could be exploited as a target for the development of novel FHB control strategies.

References

[1] Liu Y X, *et al*. Synthesis herbicidal activities and 3D-QSAR of 2-cyanoacrylates containing aromatic methylamine moieties. J Agric Food Chem, 2008, 56: 204-212.

[2] Blum G N, Nolte W A, Robertson P B. In vitro determination of the antimicrobial properties of two cyanoacrylate preparations. J Dent Res, 1975, 54: 500-503.

[3] Long N, *et al*. Synthesis and antiviral activities of cyanoacrylate derivatives containing an alpha-aminophosphonate moiety. J Agric Food Chem, 2008, 56: 5242-5246.

[4] Song B A, *et al*. Synthesis and bioactivity of 2-cyanoacrylates containing a trifluoromethyl moiety. J Fluorine Chen, 2005, 126: 87-92.

[5] Loube' ry S, Coudrier E. Myosins in the secretory pathway: tethers or transporters? Cell Mol Life Sci., 2008, 65: 2790-2800.

[6] Zhike Feng, Xiaojiao Chen, Yiqun Bao, Jiahong Dong, Zhongkai Zhang, Xiaorong Tao. Nucleocapsid of *Tomato spotted wilt tospovirus* forms mobileparticles that traffic on an actin/endoplasmic reticulum network driven by myosin XI-K.

[7] Li J F, Nebenfu hr A. The tail that wags the dog: the globular tail domain defines the function of myosin V/XI. Traffic, 2008, 116: 290-298.

[8] Sebe-Pedros A, Grau-Bove X, Richards T A, Ruiz-Trillo I. Evolution and Classification of Myosins, a Paneukaryotic Whole-Genome Approach. Genome Biol Evol, 2014, 6: 290-305.

[9] Chen L F, Bai G H, Desjardins A E. Recent advances in wheat head scabresearch in China. In: Proceedings of the International Symposium on Wheat Improvement for Scab Resistance. Suzhou and Nanjing, China, 5-11 May, 2000: 258-273.

[10] Donnell K O, Ward T J, Geiser D M, Kistler H C, Aokid T. Genealogicalconcordance between the mating type locus and seven other nuclear genes supports formal recognition of nine phylogenetically distinct species within the *Fusarium graminearum* clade. Fungal Genet. Biol., 2004, 41: 600-623.

[11] Starkey D E, Ward T J, Aoki T, Gale L R, Kistler H C, Geiser D M, Suga H, Tóth B, Varga J, O'Donnel K. Global molecular surveillance reveals novel Fusarium head blight species and trichothecene toxin diversity. Fungal Genet. Biol., 2007, 44:1191-1204.

[12] Tóth B, Mesterházy Á, Horváth Z, Bartók T, Varga M, Varga J. Geneticvariability of central European isolates of the *Fusarium graminearum* species complex. Eur. J. Plant Pathol., 2005, 113: 35-45.

[13] Nisessen L. PCR-based diagnosis and quantification of mycotoxin producing fungi. Int. J. Food Microbiol., 2007, 19: 38-46.

[14] Xu X M, Parry D W, Nicholson P, Thomsett M A, Simpson D, Edwards S G, Cooke B M, Doohan F M, Monaghan S, Moretti A, Tocco G, Mule G, Hornok L, Béki E, Tatnell J, Ritieni A. Within-field variability of Fusarium head blight pathogens and their associated mycotoxins. Eur. J. Plant Pathol., 2008, 120: 21-34.

[15] Zhang J B, Li H P, Dang F J, Qu B, Xu, Y B, Zhao C S, Liao Y C. Determination of the trichothecene mycotoxin chemotypes and associated geographical distribution and phylogenetic species of the *Fusarium graminearum* clade from China. Mycol. Res., 2007, 111: 967-975.

[16] Parry D W, Jenkinson P, McLeod L. Fusarium ear blight (scab) in small grain cereals-a review. Plant Pathol., 1995, 44: 207-238.

[17] Windels C E. Economic and social impacts of Fusarium head blight: Changing farms and rural communities in the Northern Great Plains. Phytopathology, 2000, 90: 17-21.

[18] Song B, *et al*. Type Ⅱ myosin gene in *Fusarium graminearum* is required for septation, development, mycotoxin biosynthesis and pathogenicity. Fungal Genet Biol, 2013, 54: 60-70.

[19] Sebe-Pedros A, Grau-Bove X, Richards T A, Ruiz-Trillo I. Evolution and Classification of Myosins, a Paneukaryotic Whole-Genome Approach. Genome Biol Evol, 2014, 6: 290-305.

[20] Reck-Peterson S L, Provance Jr D W, Mooseker M S, Mercer J A. 2000.

[21] Class V myosins. Biochim. Biophys. Acta. 1496: 36-51. doi:10.1016/S0167-4889(00)00007-0.

[22] Trybus K M. Myosin V from head to tail. Cell. Mol. Life Sci., 2008, 65: 1378-1389. doi:10.1007/s00018-008-7507-6.

[23] Matsui Y. Polarized distribution of intracellular components by class V myosins in *Saccharomyces cerevisiae*. Int. Rev. Cytol., 2003, 229: 1-42. doi:10.1016/S0074-7696(03)29001-X.

[24] Pruyne D, Legesse-Miller A, Gao L, Dong Y, Bretscher A. Mechanisms of polarized growth and organelle segregation in yeast. Annu. Rev. Cell Dev. Biol., 2004, 20: 559-591. doi: 10.1146/annurev.cellbio.20.01040.103108.

[25] Weisman L S. Organelles on the move: insights from yeast vacuole inheritance. Nat. Rev. Mol. Cell Biol., 2006, 7: 243-252. doi:10.1038/nrm1892.

[26] Fagarasanu A, Mast F D, Knoblach B, Rachubinski R A. Molecularmechanisms of organelle inheritance: lessons from peroxisomes in yeast. Nat. Rev. Mol. Cell Biol., 2010, 11: 644-654. doi:10.1038/nrm2960.

[27] D'ıaz-Blanco N L, Rodr'ıguez-Medina J R. Dosage rescue by UBC4 restores cell wall integrity in *Saccharomyces cerevisiae* lacking the myosin type II gene MYO1.

[28] Stevens R C, Davis T N. Mlc1p is a light chain for the unconventionalmyosin Myo2p in *Saccharomyces cerevisiae*. J. Cell Biol., 1998, 142: 711-722.

[29] Luo J, Vallen E A, Dravis C, Tcheperegine S E, Drees B, Bi E. Identifi cation and functional analysis of the essential and regulatory lightchains of the only type Ⅱ myosin Myo1p in *Saccharomyces cerevisiae*. J. Cell Biol., 2004, 165 : 843-855.

[30] Lillie S H, Brown S S. Suppression of a myosin defect by a kinesinrelatedgene. Nature, 1992, 356: 358-361.

[31] Lillie S H, Brown S S. Immunofluorescence localization of theunconventional myosin, Myo2p, and the putative kinesin-related protein,Smy1p, to the same regions of polarized growth in *Saccharomyces cerevisiae*. J. Cell Biol., 1994, 125: 825-842.

番茄晚疫病菌拮抗木霉菌株的筛选与鉴定

尉莹莹，王佳宁，梁晨*，赵洪海，李宝笃，李德龙，杨小凤
（青岛农业大学农学与植物保护学院，山东省植物病虫害综合防控重点实验室，青岛 266109）

摘　要：采用平板对峙培养法，从土壤中筛选出对番茄晚疫病菌拮抗效果较好的 5 株木霉菌株，试验结果表明 5 株木霉对番茄晚疫病菌菌株 HMQAU150020 的拮抗率为 56%～77%。结合形态学观察和 *tef*1 的系统发育分析，5 株木霉菌株分别鉴定为绿木霉（*Trichoderma virens*）、哈茨木霉（*Trichoderma harzianum*）和棘孢木霉（*Trichoderma asperellum*）。

关键词：对峙培养；木霉；番茄晚疫病菌；生物防治

Screening and identification of *Trichoderma* strains for antagonizing tomato late blight pathogen

Yingying Yu, Jianing Wang, Chen Liang*, Honghai Zhao, Baodu Li, Delong Li, Xiaofeng Yang
(Key Lab of Integrated Crop Pest Management of Shandong Province, College of Agronomy and Plant Protection, Qingdao Agricultural University, Qingdao 266109, China)

Abstract: Using confrontation test, five *Trichoderma* strains were screened from soil, and had the antagonistic effect on *Phytophthora infestans*. The experimental results indicated that the inhibited rates of five *Trichoderma* strains varied from 56% to 77%. Combining morphological observation with phylogenesis analyses of *Tef*1, these five *Trichoderma* strains were identified as *Trichoderma virens*, *Trichoderma harzianum* and *Trichoderma asperellum*, respectively.

Key words: antagonistic culture; *Trichoderma*; *Phytophthora infestans*; biocontrol

木霉（*Trichoderma* spp.）属于半知菌亚门，是一种重要的生防真菌，对多种病原菌如

基金项目：公益性行业（农业）科研专项（No. 201003004）。

作者简介：尉莹莹，女，研究生，研究方向为果蔬病害生物防治，Tel：0532-88030030，E-mail：1406134756@qq. com。
王佳宁，研究生，研究方向为蓝莓病害生物防治，Tel：0532-88030030，E-mail：1016190678@qq. com。
尉莹莹和王佳宁为并列第一作者。

* **通讯作者**：梁晨，教授，主要研究方向为真菌学及植物真菌病害的综合防治，Tel：0532-88030030，E-mail：syliangchen@163. com。

疫霉、腐霉、立枯丝核菌等具有拮抗作用[1]，随着绿色农业和有机农业的发展，以木霉为来源生防制剂的应用领域越来越广泛。

番茄是我国的重要蔬菜，由 *Phytophthora infestans* 引起的晚疫病在生产上造成严重危害。目前防治晚疫病的主要措施是化学防治，但化学药剂的长期使用造成了环境污染、农药残留及抗药性等问题。番茄生产的双减需求呼唤生物防治在晚疫病害的综合防治中发挥更积极的调控作用。本研究从多种植物的种植区采集根际土样，分离纯化并鉴定保藏木霉菌株，证明了所获得本地菌株对番茄晚疫病菌有较好的拮抗作用，为番茄晚疫病的生物防治提供了新的策略。

1 材料和方法

1.1 供试木霉菌株和病原菌

2014 年夏季于山东省各地的果蔬种植区采集根际土，采用稀释平板法[2]分离得到木霉，待长出菌落后，挑取菌落边缘单菌丝纯化，获得木霉菌株 HMQAU140012tri、HMQAU140014tri、HMQAU140015tri、HMQAU140016tri 和 HMQAU140017tri，并保存于青岛农业大学真菌学研究室。

从青岛城阳的番茄种植基地采集番茄晚疫病病样，通过单孢分离技术获得纯化的番茄晚疫病菌，致病疫霉（*Phytophthora infestans*）菌株 HMQAU150020，保存于青岛农业大学真菌学研究室。

1.2 平板对峙培养

在直径 90 mm 的黑麦培养基[3]上，在同一直线相距 4 cm 的两点上分别接入直径 5 mm 的木霉菌饼与病原菌菌饼，其中处理和对照先接入番茄晚疫病菌菌饼 4 d 后再接木霉菌饼，每个处理设三个重复，以只接病原菌菌饼的平板作为对照。20℃恒温培养箱中对峙培养。每隔 24 h 观察菌落生长情况，分别测量木霉及病原菌的生长半径。待对照病原菌长满 3/4 个皿时，按照下列公式计算抑制率。

抑制率＝（对照菌落半径－处理菌落半径）/对照菌落半径×100％

1.3 木霉菌株形态学及分子鉴定

1.3.1 木霉菌株形态学观察

挑取木霉菌落边缘菌丝制作临时玻片，在 Olympus 显微镜 BX53 下观测分生孢子梗及分生孢子的形态特征并进行显微拍照。在 25℃，12 h 光暗交替条件下培养木霉菌株，观察其在 PDA 上的菌落形态特征。参照《木霉分类与鉴定》[4]对菌株进行形态鉴定。

1.3.2 基因组 DNA 的提取

将木霉菌株接种于 PS 液体培养基中，25℃、120 r/min 摇培 48 h 后收集菌丝用于 DNA 的提取。提取方法采用改良后的 CTAB 法[5]。

1.3.3 木霉菌株 *tef*1 扩增和序列分析

提取菌株基因组 DNA 后，利用 Tef 引物 EF728（5′-CATCGAGAAGTTCGAGAAGG-3′）和 Tef1（5′-GCCATCCTTGGGAGATACCAGC-3′）进行扩增。扩增产物经 1％琼脂糖凝胶电泳检测后，由生工生物工程（上海）有限公司进行纯化和双向测序，测序结果经 Sequencher5.0 软件自动装配后导出重叠群（Contig）并在 NCBI（http://www.ncbi.nlm.gov）数据库中进行 BLAST 分析后提交给 GenBank。再从 GenBank 中选取合适的序列，经 CLUSTAL

2.0 软件进行序列对比，并用 MEGA 5.0 软件采用邻接法(neighbor-joining analysis，NJ)构建系统发育树，其中 Bootstrap 检验的重复次数为1 000次。

2 结果与分析

2.1 平板对峙培养

通过平板对峙试验观察发现，对峙培养 2 d，木霉与病原菌未接触，经测量对照病原菌半径与处理病原菌半径无差异。对峙培养 4 d，4 株木霉 HMQAU140012tri、HMQAU140015tri、HMQAU140016tri 和 HMQAU140017tri 均与病原菌已接触，并开始抑制病原菌的生长(图 1-A，B，C，D)，接触的地方形成一条拮抗线。木霉 HMQAU140014tri 虽然未与病原菌接触但病原菌靠近木霉一侧生长已经受到抑制。对峙培养 6 d，木霉已基本长满整个培养皿，病原菌停止生长，两种菌丝呈对峙状态，病原菌菌落边缘形成一条明显的拮抗线，木霉开始覆盖病原菌。(图 1-E，F，G，H)，木霉 HMQAU140014tri 还未与病原菌接触。5 株木霉对番茄晚疫病菌致疫病菌 HMQAU150020 的抑菌率均在 56% 以上，木霉 HMQAU140012tri、HMQAU140014tri、HMQAU140015tri、HMQAU140016tri 和 HMQAU140017tri 的抑制率分别为 56.82%、74.56%、77.11%、63.97% 和 64.06%，且 HMQAU140014tri 和 HMQAU140015tri 与其他 3 个菌株对病原菌的抑制率之间差异显著。

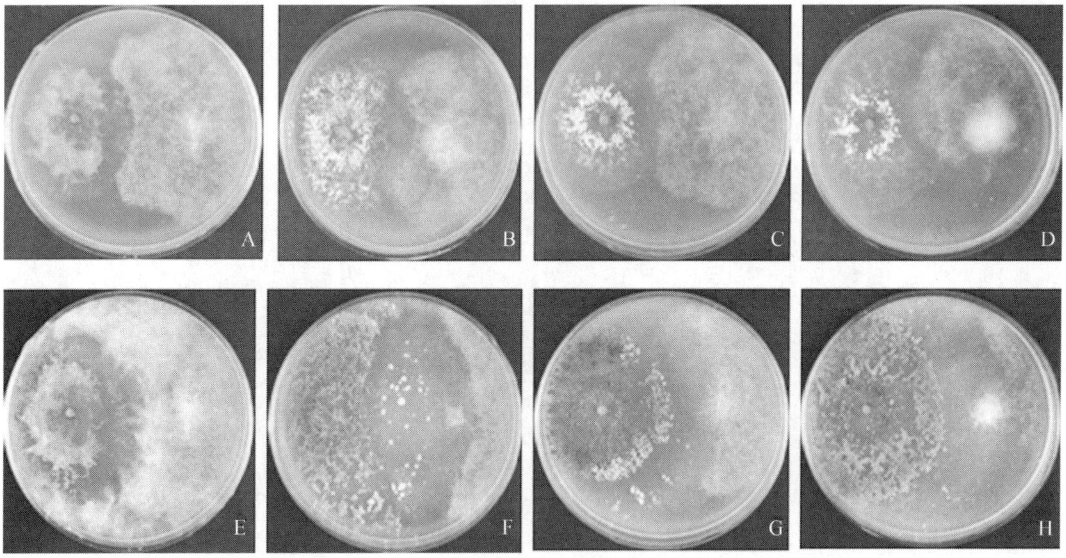

Fig. 1 Antagonistic effect of *Trichoderma* strains on tomato late blight pathogen in vitro

ABCD: Dual culture of *Phytophthora infestans* HMQAU150020 and HMQAU140012tri, 140015tri, 140016ri, 140017tri for 4 days, respectively; EFGH: Dual culture of *Phytophthora infestans* HMQAU150020 and HMQAU140012tri, 140015tri, 140016ri, 140017tri for 6 days, respectively.

2.2 木霉菌株的形态学鉴定

木霉菌株 HMQAU140012tri，在 PDA 生长速率为 17.9 mm/d，生长 3 d 可布满整个皿(直径 9 cm)。气生菌丝发达，卷毛状，白色，分生孢子在平展产孢区形成，逐渐向平板外缘扩展，后期聚集在菌落边缘，转变为黑绿色。菌落背面初期无色，后变为暗黄色，无特殊气味(图 2-A-1)。菌丝透明，厚垣孢子丰富，端生或者间生，单生，分生孢子梗半透明，分枝较为复杂，靠基部部分常常着生紧贴在一起呈漩涡状排列的分枝以及瓶梗，瓶梗基部缢缩，中部膨大，逐渐向顶端变细，长度为(9.41±2.33)μm，最宽处宽度为(3.83±0.73)μm，基部宽度为(2.46±0.42)μm，瓶梗顶端墨绿色，瓶梗上着生分生孢子(图 2-A-2)。分生孢子阔椭球形或卵圆形，两段阔圆，壁光滑，墨绿色，大小为(3.5～4.4)μm×(3.2～3.9)μm(图 2-A-3)。根据形态特征鉴定为绿木霉(*Trichoderma virens*)。

木霉菌株 HMQAU140014tri，在 PDA 上生长速率为 6.7 mm/d，气生菌丝卷毛状，不发达，白色。分生孢子产孢区形成平展的产孢簇，呈同心轮纹状分布(图 2-B-1)。分生孢子梗及分生孢子形态(图 2-B-2，3)同木霉 HMQAU140012tri。根据形态特征鉴定为绿木霉(*Trichoderma virens*)。

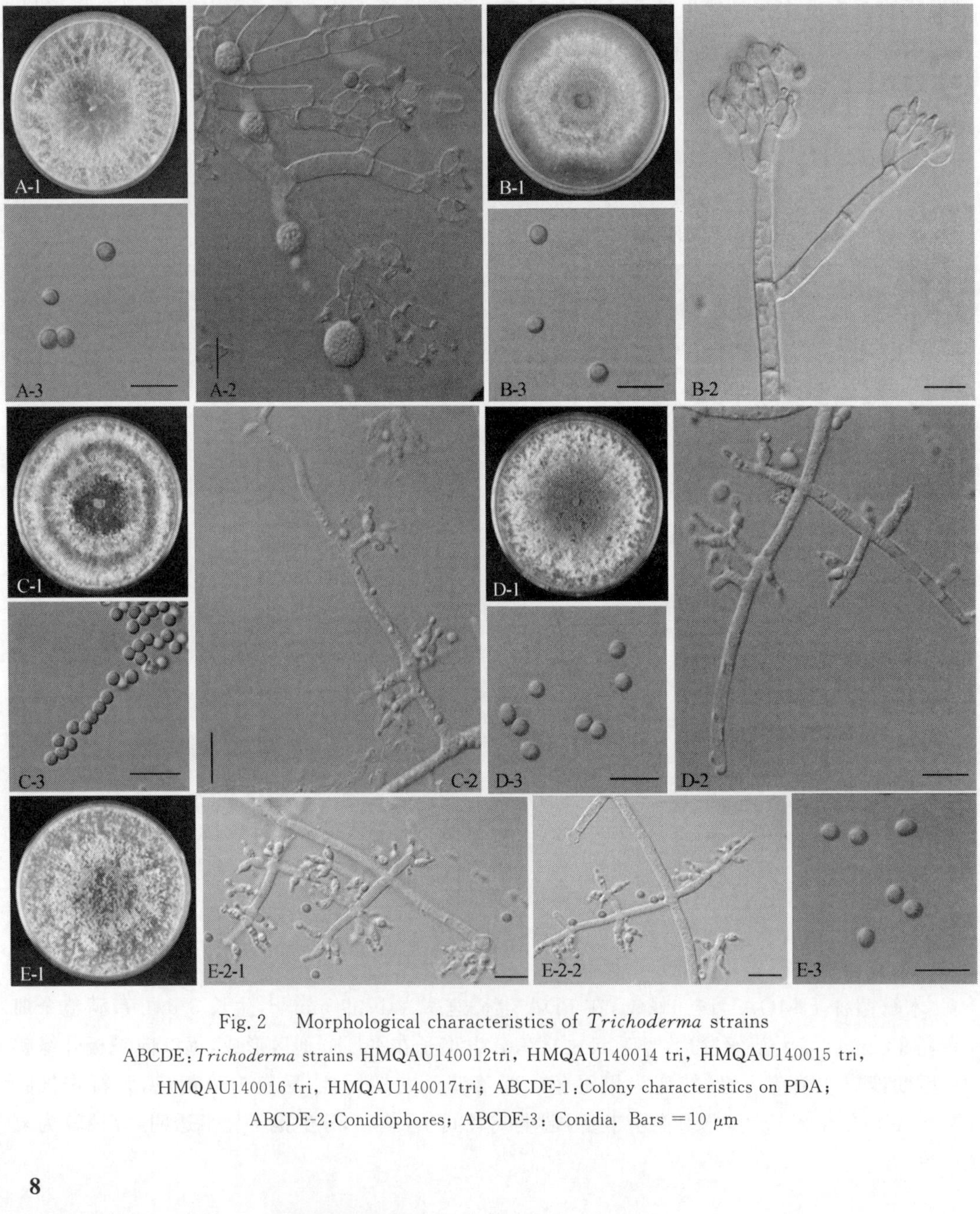

Fig. 2 Morphological characteristics of *Trichoderma* strains

ABCDE：*Trichoderma* strains HMQAU140012tri, HMQAU140014 tri, HMQAU140015 tri, HMQAU140016 tri, HMQAU140017tri; ABCDE-1：Colony characteristics on PDA; ABCDE-2：Conidiophores; ABCDE-3：Conidia. Bars =10 μm

木霉菌株 HMQAU140015tri，在 PDA 上生长速率为 14.8 mm/d，气生菌丝卷毛状，较旺盛，初期靠近中心处产生一圈分生孢子，与外圈气生菌丝呈同心轮纹状，后期气生菌丝逐渐消失，菌落边缘产生较宽的分生孢子带(图 2-C-1)。分生孢子梗分枝较复杂，初级分枝几乎呈直角状(图 2-C-2)，呈漩涡状排列，瓶梗安瓿形，长度为(6.40±0.90)μm，最宽处宽度为(3.27±0.34)μm，基部宽度为(1.87±0.25)μm。分生孢子呈卵圆形，基部细圆或者略微有尖突，大小为(3.1～3.5)μm×(2.8～3.1)μm(图 2-C-3)。根据形态特征鉴定为哈茨木霉(*Trichoderma harzianum*)。

木霉菌株 HMQAU140016tri，在 PDA 上生长速率为 12.4 mm/d，菌落初期为白色，气生菌丝卷毛状，非常发达，后期中央产生密集的分生孢子，靠近中心部位产孢区黑绿色，缺乏气生菌丝，产孢区逐渐形成同心轮纹状，菌落背面有褶皱结构(图 2-D-1)。分生孢子梗主轴顶端下面生出的初次分枝常呈对生，与主轴夹角近 90°，瓶梗呈对称分布，长度为(9.09±1.79)μm，最宽处宽度为(3.14±0.61)μm，基部宽度为(1.99±0.31)μm(图 2-D-2)。分生孢子球形或卵圆形，大小为(3.8～4.5)μm×(3.2～3.7)μm(图 2-D-3)。根据形态特征鉴定为棘孢木霉(*Trichoderma asperellum*)。

木霉菌株 HMQAU140017tri，菌落与 HMQAU140016tri(图 2-E-1)较相似，在 PDA 上生长速率为 15.8 mm/d，初期气生菌丝相对 140016 略少，后期形成同心轮纹状产孢区不是特别明显。分生孢子梗(图 2-E-2)及分生孢子(图 2-E-3)同 HMQAU140016tri。根据形态特征鉴定为棘孢木霉(*Trichoderma asperellum*)。

2.3 木霉菌株的分子鉴定和系统发育分析

将 5 株木霉的 *tef*1 序列提交至 GenBank 数据库中，进行 BLAST 相似性比对，结果表明菌株 HMQAU140012tri 和 HMQAU140014tri 与绿木霉(*Trichoderma virens*，GenBank 登录号为 JQ040420)的同源性达到 99%，将两个菌株的序列提交到 GenBank，登录号分别为 KP747446 和 KP747447。菌株 HMQAU140016tri 和 HMQAU140017tri 与棘孢木霉(*Trichoderma asperellum*，GenBank 登录号为 KM190855 和 JQ040488)的同源性均达到 99%，将两个菌株的序列提交到 GenBank，登录号分别为 KP747448 和 KP747449。菌株 HMQAU 140015tri 与哈茨木霉(*Trichoderma harzianum*，GenBank 登录号为 HQ222309)的同源性达到 99%，将该菌株的序列提交到 GenBank，登录号为 KP747450(表 1)。根据 *tef*1 基因系统发育分析结果，菌株 HMQAU140012tri 和 HMQAU140014tri 与绿木霉位于系统发育树的同一分支；菌株 HMQAU140016tri 和 HMQAU140017tri 序列与棘孢木霉位于系统发育树的同一分支；HMQAU140015tri 与哈茨木霉位于系统发育树的同一分支(图 3)。同源性比对数据和系统发育树位置进一步证明了菌株 HMQAU140012tri 和 HMQAU140014tri 为绿木霉(*Trichoderma virens*)，菌株 HMQAU140016tri 和 HMQAU140017tri 为棘孢木霉(*Trichoderma asperellum*)，HMQAU140015tri 为哈茨木霉(*Trichoderma harzianum*)。

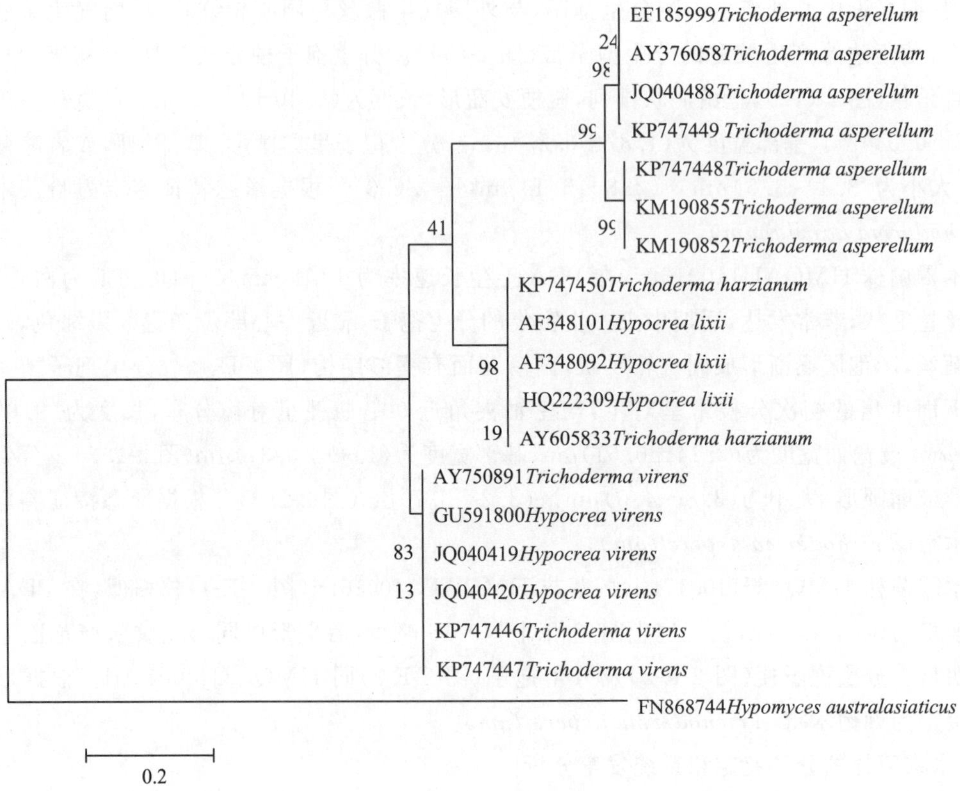

Fig. 3　Neighbour joining phylogenetic tree based on *tef*1 gene sequences of *Trichoderma* spp. and related fungi

Numbers at nodes indicating bootstrap values for each node out of 1 000 bootstrap replications;
Scale 0.1 meaning evolutional distance. Outgroup: *Hypomyces australasiaticus* FN868744

Table1　The *tef*1 sequence accession number of related *Trichoderma* strains in neighbor-joining phylogenetic analyses

种类 Species	来源 Origin	寄主 Host	GenBank 登录号 GenBank Accession No.
Trichoderma asperellum	喀麦隆 Cameroon	根际土壤 Rhizosphere soil	EF185999
Trichoderma asperellum	美国 USA	小核盘菌 *Sclerotinia minor*	AY376058
Trichoderma asperellum	中国，上海 Shanghai, China	土壤 Soil	JQ040488
Trichoderma asperellum	中国，山东 Shandong, China	土壤 Soil	KP747449
Trichoderma asperellum	中国，山东 Shandong, China	土壤 Soil	KP747448
Trichoderma asperellum	印度 India	土壤 Soil	KM190855
Trichoderma asperellum	印度 India	土壤 Soil	KM190852
Trichoderma harzianum	中国，山东 Shandong, China	土壤 Soil	KP747450

续 Table1

种类 Species	来源 Origin	寄主 Host	GenBank 登录号 GenBank Accession No.
Hypocrea lixii	美国 USA	双孢菇 *Agaricus bisporus*	AF348101
Hypocrea lixii	美国 USA	双孢菇 *Agaricus bisporus*	AF348092
Hypocrea lixii	中国，上海 Shanghai, China	土壤 Soil	HQ222309
Trichoderma harzianum	未知 Unknown	未知 Unknown	AY605833
Trichoderma virens	美国，乔治亚州 Georgia, USA	土壤 Soil	AY750891
Hypocrea virens	中国，上海 Shanghai, China	土壤 Soil	JQ040419
Hypocrea virens	美国，佐治亚州 Georgia, USA	土壤 Soil	GU591800
Hypocrea virens	中国，上海 Shanghai, China	土壤 Soil	JQ040420
Trichoderma virens	中国，山东 Shandong, China	土壤 Soil	KP747446
Trichoderma virens	中国，山东 Shandong, China	土壤 Soil	KP747447
Hypomyces australasiaticus	泰国 Thailand	担子菌 *Basidiomycetes*	FN868744.1

3　结论与讨论

通过平板对峙培养，筛选出对番茄晚疫病菌 HMQAU150020 具有较高拮抗作用的 5 株木霉菌株，5 株木霉对番茄晚疫病菌的拮抗率为 56%～77%。经形态学观察和 *tef*1 序列的系统发育分析，将 5 株木霉菌株分别确定为绿木霉(*Trichoderma virens*)、哈茨木霉(*Trichoderma harzianum*)、棘孢木霉(*Trichoderma asperellum*)。

对峙培养试验中发现木霉菌株 HMQAU140015tri 对病原菌均表现出较好的拮抗效果，这可能与其较快的生长速率(图 1)有关，该菌株能够快速覆盖病原菌，抑制病原菌生长。另一木霉菌株 HMQAU140014tri 生长速率最慢，但在菌株与病原菌接触之前病原菌的生长就开始受到抑制，并形成一条明显的拮抗线。推断出木霉 HMQAU140014tri 可能分泌某种抗生素，而且该抗生素对病原菌有较好的拮抗效果。木霉对病原真菌的拮抗作用主要包括以下机制：重寄生作用，产生抗生素、酶类物质，竞争作用，诱导抗性，促进植物生长，协调植物的抗性等[6]。对于这两个木霉菌株的生防机制需要进一步探索。

而 Chaverri(2001)曾采用 ITS 和 *tef*1 序列分析绿木霉(*Trichoderma virens*)的有性阶段，结果发现采用 ITS 序列分析五种绿木霉聚在同一分支的支持率为 100%，而 *tef*1 基因序列分析显示出了 5 株木霉之间的差异，因此推断 *tef*1 基因要比 ITS 基因更适用于木霉的多基因序列分析[8]。Dodd 等(2000)曾研究表明，采用 ITS 分析技术描述木霉进化枝，有时候并不足以确认关系密切分类单元之间的进化关系问题[9]。Kullning-Gradinger 等(2002) 综合利用 ITS、28S、mitSSU、*tef*1、和 *ech*42 序列用于木霉属的系统发育分析[11]。因此，对于多株木霉的系统发育问题应该结合更多种类的序列分析方法。

另外，木霉菌株的准确鉴定将为其生防机理和应用研究奠定了基础。Chaverri(2001)曾采用 ITS 和 *tef*1 序列分析绿木霉 *Trichoderma virens* 的有性阶段，结果发现采用 ITS 序列分析五种绿木霉聚在同一分支的支持率为 100%，而 *tef*1 基因序列分析显示出了 5 株木霉之间的差异，因此推断 *tef*1 基因要比 ITS 基因更适用于木霉的多基因序列分析[7]。Dodd 等(2000)曾研究表明，采用 ITS 分析技术描述木霉进化枝，有时候并不足以确认关系密切分类单元之间的进化关系问题[8]。而且 Chaverri(2001)还发现根据 ITS 和 *tef*1 的系统发育分析，绿木霉与哈茨木霉亲缘关系较近[7]。本研究曾利用 ITS 基因序列分析 5 株木霉的分类地位及进化关系，结果与形态学鉴定及 *tef*1 基因序列存在分歧；因此本试验采用 *tef*1 基因序列发育分析以确定 5 株木霉的分类地位及进化关系。研究发现无论是 ITS 还是 *tef*1 系统发育分析，HMQAU140015tri 与绿木霉 HMQAU140012tri 和 HMQAU140014tri 亲缘关系较近。加之，根据形态学观察木霉菌株 HMQAU140015tri 应属于哈茨木霉，因此结合形态学和基因序列分析木霉菌株 HMQAU140015tri 应属于哈茨木霉（*Trichoderma harzianum*）。Kullning-Gradinger 等(2002) 综合利用 ITS、28S、mitSSU、*tef*1 及 *ech*42 序列用于木霉属的系统发育分析[9]。因此，今后应该结合更多种类的序列分析方法。

传统的木霉菌株鉴定主要依据形态学特征，虽然木霉很容易辨认，而且分类研究也比较深入[10]，但木霉的形态性状复杂多变且易受环境影响，缺少特有形态特征，给分类研究造成了极大的不便，也是木霉分类存在较大分歧的原因之一[11]。近年来，木霉的鉴定逐渐过渡到分子手段，分类性状从比较单一的形态学性状发展到综合形态学、生理生化、分子系统学等各方面的性状[12]。随着分子生物技术的发展，木霉的分类系统必将更加完善，形成一套比较完备的分类鉴定体系。

参考文献

[1] Papavizas G C. Biological control of selected soilborne plant pathogens with *Gliocladium* and *Trichoderma*[J]. Biological Control of Plant Diseases, 1992, 230: 223-230.

[2] Fang Z D. Methodology on plant pathology research (in Chinese) [M](3rd ed.). Beijing: China Agricultural Press, 1998, 132-133.

[3] Zheng X B. Genus *Phytophthora* and research method of *Phytophthora*[M]. Beijing: Chinese Agricultural Press, 1995, 81-82.

[4] Yang H T, Classification and identification of *Trichoderma*[M]. Beijing: China Land Press, 2009, 1-364.

[5] Zhang Y H, Wei D S, Xing L J, *et al* . A modified method for isolating DNA from fungus[J]. Microbiology(微生物学通报), 2008, 35(3): 466-469.

[6] Howell C R, Stipanovic R D. Gliovirin, a new antibiotic from *Gliocladium virens*, and its role in the biological control of *Pythium ultimum*[J]. Canadian Journal of Microbiology, 1983, 29(3): 321-324.

[7] Chaverri P, Samuels G J, Stewart E L. *Hypocrea virens* sp. nov. the teleomorph of *Trichoderma virens*[J]. Mycologia, 2001, 93: 1113-1124.

[8] Dodd S L, Crowhurst R N, Rodrigo A G, *et al*. Examination of *Trichoderma* phylogenies derived from ribosomal DNA sequence date[J]. Mycological Research, 2000, 104: 23-34.

[9] Kullnig-Gradinger C M, Szakacs G, Kubicek C P. Phylogeny and evolution of the genus *Trichoderma*: a multigene approach[J]. Mycological Research, 2002, 106: 757-767.

[10] Gams W, Bissett J. Gams W, Bissett J. Morphology and identification of *Trichoderma* [A]. *Trichoderma* and *Giocladium*: Basic biology, taxonomy and genetics. CRC Press: Harman GE, Kubicek C P, 1998. 1: 3-56

[11] Bissett J. A revision of the genus *Trichoderma*. Ⅳ. Additional notes on section *Longibrachiatum* [J]. Canadian Journal of Botany, 1991, 69(11): 2418-2420.

[12] Zhang C L, Xu T. Advances on molecular phylogeny and classification of the genus *Trichoderma* and its related teleomorphs[J]. Journal of Zhejiang University (浙江大学学报), 2003, 29(4): 461-470.

河北省玉米根腐病病原菌组成及防治药剂筛选

纪莉景，栗秋生，王连生，李聪聪，肖颖，孔令晓*

（河北省农林科学院植物保护研究所，农业部华北北部作物有害生物综合治理重点实验室，河北省有害生物综合防治工程技术研究中心，保定 071000）

摘　要：对河北省玉米根腐病病原菌组成进行研究并对玉米根腐病菌有防治效果的药剂进行筛选，结果表明，引起河北省玉米根腐病的主要病原菌为藤仓赤霉复合种、离蠕孢菌、禾谷镰刀菌、丝核菌和腐霉菌，并且不同地区主要病原菌的种类存在差异；对主要病原菌藤仓赤霉复合种、离蠕孢菌、禾谷镰刀菌进行有效防治药剂筛选，满适金对藤仓赤霉复合种防治效果较好，防效达到 59.7％，3％敌委丹＋2％立克秀组合对离蠕孢菌防治效果较好，防治效果达到 29.4％，适麦丹对禾谷镰刀菌防治效果较好，防效为 50.3％。

关键词：玉米根腐病；藤仓赤霉复合种；离蠕孢菌；药剂筛选

Pathogen composition and fungicide screening for maize root rot diseases in Hebei

Lijing Ji, Qiusheng Li, Liansheng Wang, Congcong Li, Ying Xiao, Lingxiao Kong*

(Key Laboratory of Integrated Pest Management on Crops in Northern Region of North China, Ministry of Agriculture; Integrated Pest Management Center of Hebei Province, Institute of Plant Protection, Hebei Academy of Agricultural and Forestry Sciences, Baoding 071000, China)

Abstract: Pathogen composition and fungicide screening for maize root rot diseases were conducted in Hebei province. Results showed that the main pathogens associated with maize root rot disease were *Gibberella fujikuroi* species complex, *Biplaris sorokiniana*, *Fusarium graminearum*, *Rizoctonia* spp. and *Pythium* spp.. In addition, the predominant pathogen was different among regions. Fungicide screening on *Gibberella fujikuroi* species complex, *B. sorokiniana* and *F. graminearum* showed that fludioxonil＋metalaxyl, 3％ difenoconazole ＋2％ tebuconazole and fludioxonil＋difenoconazole had better control effect on *Gibberella fujikuroi* species complex, *B. sorokiniana* and *F. graminearum* respectively, with control efficacy 59.7％, 29.4％ and 50.3％.

基金项目：国家科技支撑计划（2012BAD19B04-08）；河北省科技支撑计划（13226509D）。

* **通讯作者**：孔令晓，研究员，主要从事植物病害研究，E-mail：konglingxiao163@163.com。

Key words: maize root rot diseases; *Gibberella fujikuroi* species complex; *Biplaris sorokiniana*; fungicide screening

玉米根腐病是由多种病原菌引起的，主要表现为中胚轴和整个根系逐渐变褐腐烂，根系生长受阻，植株矮小，叶片发黄，严重时引起幼苗死亡，玉米根腐病主要在苗期危害初生根，对气生根和后期植株生长影响不大。近年来，玉米根腐病在各玉米产区的发生有逐年加重的趋势，重病地块病株率高达100%，严重时全田死苗，造成毁种，成为玉米生产上的新障碍[1,2]。玉米根腐病主要由镰刀菌、丝核菌、腐霉菌和蠕孢菌等引起的[3,4]，但不同生态区域优势致病菌的种类存在差别。李建强等曾报道，种子包衣能够提高玉米出苗率并对玉米粗缩病和苗枯病有较好的防治效果[5]。然而实行新的耕作制度以来，河北省玉米根腐病病原菌的组成和地区分布鲜见报道，生产上也缺乏针对不同地区病原菌组成差异相应的防治措施，本文对河北省玉米根腐病的病原菌组成和地区间分布以及对主要病原菌有防治效果的药剂进行筛选，为针对地域间不同病原菌组成玉米根腐病相应防治措施的制定提供基础。

1 材料与方法

1.1 河北省玉米根腐病标样的采集和病原菌分离鉴定

2008—2013年7月上中旬，分别在河北省夏玉米主产区石家庄、保定、沧州、衡水、邢台等地采集3～4叶期玉米苗标样，采集田块前茬作物多为小麦，每块田块采集40株左右，取样时将植株连根挖出，尽量保持玉米根系完整，随即带土装入贴好标签的塑料袋，带回实验室。将采集的各病株的根部剪下，流水冲洗12 h，用75%乙醇表面消毒，每株病根切取3段放到马铃薯葡萄糖琼脂培养基(PDA)上，25℃培养，待菌落长出后，根据菌落形态、孢子类型鉴定各病株真菌培养物的种类[6,7]，统计各地区玉米根腐病病原菌的组成及分离频率。

1.2 防治玉米根腐病有效种衣剂的筛选

选取分离频率较高的藤仓赤霉复合种、离蠕孢菌和禾谷镰刀菌菌株进行7种有效种衣剂及组合的筛选，所选用药剂分别为2%立克秀、4.8%适麦丹、3%敌委丹、3.5%满适金、2.5%适乐时、6.25%亮盾和3%敌委丹+2%立克秀组合。采用土壤带菌法接种，于PDA平面上扩繁待接种菌株，用直径0.6 cm打孔器打成菌饼，移植于三角瓶内的谷粒培养基上培养15 d左右，每隔一天摇动一次，待谷粒布满菌丝后，将培养物与灭菌土(沙子：蛭石=1：1)以体积比1：5混合均匀，装入直径13 cm的花盆中。按各种衣剂推荐浓度进行种子包衣，所用玉米品种郑单958，每种衣剂每病原菌处理设5次重复，每重复播种10粒种子，设不包衣的为空白对照，播种后覆灭菌土，置于25～28℃温室内常规培养，3周后，按照0～9级分级方法调查每个植株的发病级别，计算病情指数和防治效果。

玉米苗期根部调查分级标准如下：

0:无症状；

1:根部零星发病，病根面积占整个根面积的5%以下；

3:病根面积占根面积的5%～25%；

5:病根面积占根面积的26%～50%；

7:病根面积占根面积的 51%～75%；

9:病根面积占根面积的 75%以上,几乎整株根部变褐。

病情指数=100× $\sum$(各级病株数×各级代表值)/(调查总株数×最高级代表值)

2 结果与分析

2.1 河北省玉米根腐病病原菌组成和地区间分布

2008—2013 年对采集的 1 427 株玉米根腐病病株进行病原菌分离,分离到 5 种主要病原菌分别为藤仓赤霉复合种(原串珠镰刀菌)、离蠕孢菌、禾谷镰刀菌、丝核菌和腐霉菌,其中,分离频率最高的为藤仓赤霉复合种占 39.7%,其次为离蠕孢菌和禾谷镰刀菌,分离频率分别为 16.6%和 2.9%。

不同地区各优势病原菌的分离频率不同,在石家庄、保定、沧州的玉米田中,以藤仓赤霉复合种的分离频率最高,为 35.2%～50.6%,而在衡水和邢台地区玉米标样病原菌分离中,以离蠕孢菌分离频率偏高,分别为 25.0%和 24.3%,禾谷镰刀菌在沧州和衡水的分离频率较其他地方偏高。由此可见,藤仓赤霉复合种、离蠕孢菌和禾谷镰刀菌为引起河北省玉米根腐病的主要病原菌。

2.2 防治玉米根腐病有效种衣剂的筛选

7 种种衣剂及组合包衣对玉米根腐病防治效果盆栽试验结果表明,在供试浓度下对禾谷镰刀菌引起的玉米根腐病防治效果较好的药剂为适麦丹,防效为 50.3%,其次为满适金和适乐时,防效分别为 42.6%和 42.2%;对藤仓赤霉复合种引起的玉米根腐病防治效果较好的为满适金,防效达到 59.7%,其次为立克秀防效为 47.1%;对离蠕孢菌引起的玉米根腐病防治效果较好的药剂为 3%敌委丹+2%立克秀组合,防治效果达到 29.4%,其次为适麦丹和敌委丹,防效分别为 27.8%和 20.6%。

3 讨论

玉米根腐病是我国玉米产区普遍存在、发生程度逐年加重的一类土传病害,据报道,玉米根腐病主要由串珠镰刀菌、禾谷镰刀菌、立枯丝核菌、腐霉菌和蠕孢菌等多种病原菌单一或复合侵染引起的[1,2],在本研究中河北省玉米根腐病是由多种病原菌复合侵染造成的,而且病原菌的种类与以前报道的基本一致。但是值得注意的是,河北省各地区玉米根腐病病害症状相似,但地区间病原菌优势种存在差异,由于各病原菌间致病力存在差别,因此,各地区由不同优势病原菌造成的病害损失程度可能不同[8],调查采样过程中虽然未见玉米根腐病引起大面积死苗现象,但在所采样地块发生普遍,且严重病株已经影响了玉米根系和植株生长,而发病较轻的植株,玉米次生根和须根快速生长可以弥补种子根受损造成的影响,对后期玉米生产不会造成很大的危害,因此,在发病严重的地块应当提前采取种子处理或玉米苗期预防措施,以防止根腐病对玉米生产造成损失。此外,本研究结果显示不同种衣剂包衣对不同病原菌的防治效果存在差异,因此在病害防治中应当充分了解地区间致病菌的组成,因地制宜地制定相关防治措施和合理用药,以使病害防治达到事半功倍的效果。

参考文献

[1] 孙广勤. 鲁西夏玉米苗期叶枯病发生特点与防治对策. 作物杂志,2003(4):43-44.

[2] 李军,赵雪峰,邓如正. 玉米苗期根腐病的发生与防治. 农业开发与装备,2014(7):123.

[3] 赵晓军,石秀清,赵子俊,等. 武将 450FS 悬浮种衣剂对玉米苗期病害的防治效果. 中国种业,2006(8):29-30.

[4] 李红,晋齐鸣,王立新. 吉林省玉米苗期病虫害发生动态及防治对策. 吉林农业科学,2002,27(增刊):33-34.

[5] 李建强,李洪连,袁红霞,等. 种子包衣防治玉米苗期病害及对生长和产量的影响. 中国农业大学学报,1999, 4(5): 82-86.

[6] 魏景超. 真菌鉴定手册. 上海:上海科学技术出版社, 1979:1-802.

[7] Leslie J F, Summerell B A. The *Fusarium* laboratory manual. Iowa: Blackwell Publishing, 2006:1-388.

[8] 纪莉景,栗秋生,王连生,等. 河北省夏玉米苗期根病发生现状及病原初探. 玉米科学, 2014, 22(6): 138-141.

苹果斑点落叶病不同药剂防治效果的比较研究

郝婕[1]，魏亮[2*]，王献革[1*]，李学营[1]，索相敏[1]，鄢新民[1]，冯建忠[1]
([1]河北省农林科学院石家庄果树研究所，石家庄 050061；
[2]河北省国土资源厅，石家庄 050051)

摘　要：对防治苹果斑点落叶病的常见3种药剂进行筛选，通过比较3种药剂的相对防治效果，确定了针对该病的最佳防治药剂及适宜浓度。结果表明，50%扑海因1 000～1 500倍液可湿性粉剂的防效最佳，其次是80%代森锰锌600～800倍液。本研究可为指导苹果树科学施药提供理论依据。

关键词：苹果；斑点落叶病；防治效果

Comparative studies on apple *Alternaria mali* control efficiency by different medicament treatment

Jie Hao[1], Liang Wei[2*], Xiange Wang[1*], Xueying Li[1],
Xiangmin Suo[1], Xinmin Yan[1], Jianzhong Feng[1]
([1]Shijiazhuang Pomology Institute, Hebei Academy of Agriculture and Forestry Sciences, Shijiazhuang 050061; [2]Hebei Province Department Land and Resources, Shijiazhuang 050051)

Abstract: The objective of this study is to determine the best medicament and suitable concentration on the relative control effect for apple *Alternaria mali*, which selected by four common medicaments. The results showed that 50% Iprodione for 1 000-1 500 times liquid of the wettable powder had the best control effect, while 80% Mancozeb for 600～800 times liquid of the wettable powder followed. The results provided scientific theoretical basis for guiding the apple tree pesticide application scientifically.

Key words: apple; *Alternaria mali*; control efficiency

斑点落叶病是苹果生产上常见的主要病害之一，属半知菌亚门真菌[1]。其病菌以菌丝体和分生孢子盘在病叶上越冬，第二年春天产生分生孢子，通过风雨传播，直接或从气孔侵染。以叶龄20 d内的嫩叶易受侵染，30 d以上叶不再感病。该病潜育期短，一般6～12 d，

作者简介：郝婕(1979—　)，女，河北石家庄人，副研究员，硕士，研究方向为果树育种及病虫害防控研究，E-mail：haohao_822@163.com。

*** 通讯作者**：魏亮(1982—　)，男，河北赵县人，工程师。王献革(1966—　)，女，河北栾城人，副研究员。Tel：0311-87659934，E-mail：guoshusuofjz@126.com。

在田间具有多次再侵染。病菌从侵染到引起落叶需13～55 d。田间一般从5月上中旬开始发病，6～9月份为发病盛期，严重时9月份即可造成大量落叶。近几年，该病在河北省中南部苹果主产区有大面积发生的现象，造成当地果园苹果植株叶片大量脱落，树体长势衰弱，直接影响当年果树花芽分化及果实发育，且对来年果树的长势和产量也造成很大影响[1-3]。品种以红星、玫瑰红、元帅系列易感病；富士系列、乔纳金、鸡冠等发生较轻[3]。本文通过不同药效试验，筛选出适宜的药剂种类及适宜浓度，从而为指导科学施药提供切实可行的实验基础和科学依据。

1 材料与方法

1.1 材料

试验于2013年在河北省深县清辉头村（河北中南部苹果产区）苹果示范园进行，选择地势低洼、地下水位较高、发病较重的果树区域进行。

试验品种为新红星12年生，株行距为2 m×3 m。试验株选用生长势基本一致的新红星苹果树。

1.2 试验方法

1.2.1 试验处理

选定调查对象，并做标记，以单株为一小区。试验设4个药剂处理：1.5%多抗霉素400～500倍液、50%扑海因可湿性粉剂1 000～1 500倍液、80%代森锰锌可湿性粉剂600～800倍、清水对照（CK）。每个处理重复3次，共15个小区，随机排列。

用踏板式高压喷雾器均匀、淋洗状喷施。处理时间和间隔：5月22日喷施第1次，间隔15 d喷第2次，再间隔15 d后喷第3次。

1.2.2 试验调查

于5月22日、6月24日、7月23日、8月24日、9月24日、10月24日分6次调查各处理叶片感病指数。每株树按东、南、西、北四个方位，每个方位各随机50片叶（包括外围、内膛和长枝、短枝各处的叶片），共200片叶进行调查。

1.2.3 计算方法[3-6]

将3种不同药剂及清水对照组处理的新红星苹果叶片进行采集和统计，按照下述方法划分病级，并计算各药剂处理下叶片的病叶率、病情指数及相对防治效果。

病级划分标准：0级，叶片无病斑；1级，叶片有零星小病斑；2级，叶片病斑面积占叶面积的1/4；3级，叶片病斑面积占叶面积的1/3；4级，叶片病斑面积占叶面积的1/2以上。

病叶率（%）的计算方法：病叶片数/总叶片数×100%

病情指数（%）的计算方法为：$\sum$（各级病叶数×相应各级代表值）/（调查总叶数×最高级代表值）×100%

相对防治效果（%）的计算办法为：（1－防治区病情指数/对照区病情指数）×100%

2 结果与分析

2.1 不同药剂处理对新红星苹果斑点落叶病的防治效果

试验期间，果园温、湿度适中。50%扑海因1 000～1 500倍液自病害发生后的相对防治效果最为显著，在病害发生的中、后期施药效果最佳，可达到92.7%；80%代森锰锌600～

800 倍液的防效次之，在病害发生后喷药初期即能表现出防治效果，其最高防效可达到 90.6%；1.5%多抗霉素 400～500 倍液在病害发生的前期未表现出防治效果，只在后期才表现出一定的防治效果。见表 1。

表 1　不同药剂处理对新红星苹果斑点落叶病防治效果

Table 1　Control efficiency on Starkrimson apple Alternaria mali Roberts by different medicament treatment

不同药剂处理类型	调查内容	5 月 22 日	6 月 24 日	7 月 23 日	8 月 24 日	9 月 24 日	10 月 24 日
1.5%多抗霉素 400～500 倍液	病叶率(%)	10.6	15.0	15.6	3.3	5.0	5.0
	病情指数(%)	5.0	7.6	8.2	2.1	2.3	2.6
	相对防效(%)	—	—	37.3	88.8	67.2	72.4
50%扑海因 1 000～1 500 倍液	病叶率(%)	7.2	12.8	22.2	5.6	1.1	1.1
	病情指数(%)	3.6	6.4	8.5	2.2	0.7	0.7
	相对防效(%)	—	—	35.4	88.1	89.6	92.7
80%代森锰锌 600～800 倍液	病叶率(%)	2.8	2.8	19.4	3.3	1.1	3.9
	病情指数(%)	1.3	1.4	7.8	1.9	0.7	2.0
	相对防效(%)	64.8	67.8	40.8	89.7	90.6	79.2
清水对照	病叶率(%)	6.1	7.8	35.0	37.2	11.1	18.9
	病情指数(%)	3.6	4.4	13.2	18.9	6.9	9.5

2.2　不同药剂处理对苹果斑点落叶病防治的周年变化规律

对各药剂处理组下苹果斑点落叶病叶片的相对防治效果进行周期性跟踪调查，自 5 月 22 日开始，分别在 6 月 24 日、7 月 23 日、8 月 24 日、9 月 24 日、10 月 24 日采集受害叶片，计算感病指数，根据病害的相对防治效果计算公式得到各时期叶片的相对防治效果，制作图 1。由图 1 可看出各药剂处理对苹果斑点落叶病均有不同程度的防治效果。50%扑海因 1 000～1 500 倍液的防治效果在 5 月份未表现出有防治作用，而自 6 月份开始，以后均呈极显著的、连续的上升态势，至 10 月份达到最高，防治效果最为突出。80%代森锰锌 600～800 倍液作用下相对防治效果在整体上均呈一定防治效果，早期就呈现出 60%以上的相对防效，自 7 月份以后开始呈较明显的上升规律，整体防治效果较为突出。1.5%多抗霉素 400～500 倍液前期未有防治效果，1.5%多抗霉素 400～500 倍液在 7 月份之后的相对防效迅速升高，8 月份达到最高，随后略有下降波动，可作为后期防治斑点落叶病药剂的选择种类。

3　结论与讨论

苹果斑点落叶病是我国苹果产区的重要病害之一，该病发病普遍，危害严重[4,6,9]。其主要危害叶片，也可侵染果实和叶柄，导致早期落叶。叶片染病初期出现褐色圆点，其后逐渐扩大为红褐色，边缘紫褐色，病部中央常具一深色小点或同心轮纹。其病菌以菌丝团或分生孢子盘在落叶上越冬，第二年春季降雨病部中央常具一深色小点或同心轮纹。其病菌以菌丝团或分生孢子盘在落叶上越冬，第二年春季降雨后萌发分生孢子，6～9 月份为发病盛期。其发病轻重与 4～5 月份降雨量及田间管理措施关系密切，雨水是病菌传播和病害流行的主导因素，凡是春季多雨、夏秋雨季提前、高温潮湿的年份，病害就会大流行，8～9 月份造成易感病品种叶片大量脱落。此外，树势和品种也与苹果斑点落叶病密切相关，凡是地势低洼、树冠郁闭、通风透光条件差、管理粗放、树势衰弱的果园均易感病，感病品种以新红星、玫瑰红、元帅系等品种的发病较重[5-7]。本试验针对冀中南地区斑点落叶病又有大面积发生的现状，将生产上常用的 3 种药剂进行了周年药效的跟踪试验和相对防治效果的科学筛选，可为果农选择最佳防治药剂及合理喷布浓度提供了科学参考和借鉴。经本试验研究表明，在斑点落叶病的发病初期，可选择用 80％代森锰锌 600～800 倍液进行防治，在发病中、后期则以 50％扑海因 1 000～1 500 倍液为主，80％代森锰锌 600～800 倍液、1. 5％多抗霉素 400～500 倍液为辅，交替施药，能有效控制苹果斑点落叶病的发生。从单剂的相对防效上看，50％扑海因 1 000～1 500 倍液防治最佳，年防治次数以 5～7 次为宜。

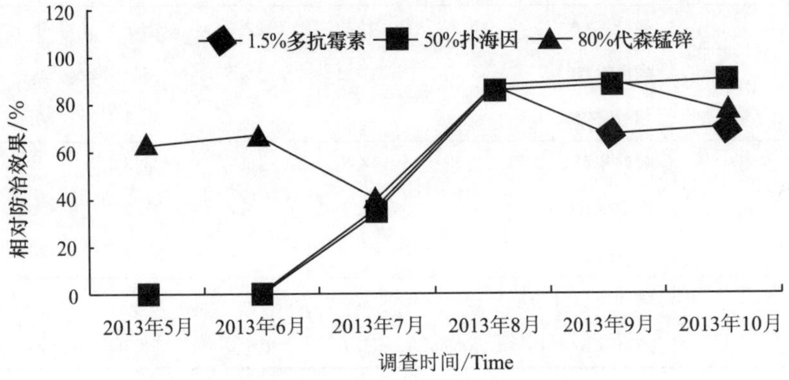

图 1　苹果斑点落叶病不同药剂相对防治效果的周年变化规律

Fig. 1　Annual change regularity of Starkrimson apple *Alternaria mali* by different medicament treatment

针对苹果斑点落叶病，除采用有效药剂进行联合防治外，还应配合加强田间管理措施，增势有机肥和磷钾肥，控制氮肥使用[8,9]；合理修剪，加强夏剪，保证树冠通风透光；秋冬季清理果园残枝落叶，进行树干涂白和 5 波美度石硫合剂等，提高树体自身免疫力，减少病原菌流行[9,10]。

参考文献

[1]　Duan J F. Efficacy of several fungicides in controlling apple leaf spot and apple ring rot(in Chinese)[D]. Baoding：Agricultural University of Hebei(保定：河北农业大学)，2012.

[2]　Mohammad J S, Marzieh E. First report of *Alternaria mali* causing apple leaf blotch disease in Iran[J]. Australasian Plant Disease Notes，2007，2：57-58.

[3]　Abe K，Iwanami H，Kotada N，*et al*. Evaluation of apple genotypes and Malus species for resistance to Alternaria blotch caused by *Alternaria alternata* apple pathotype using detached-leaf method[J]. Plant Breeding，2010，129：208-218.

[4]　Zhang X. The toxicity determination and field application of tebuconazole 50％ nano spx on the *Alternaria alternata* f. sp. *mali* (in Chinese)[D]. Yangling：Northwest A&F University(杨凌：西北农林科技大学)，2014.

[5]　Wei J G. The toxicity determination and field application of difenoconazole azole mixed with polyoxin different combinations on the *Alternria alternata* f. sp. *mali*(in Chinese)[D]. Yangling：Northwest A&F University(杨凌：西北

农林科技大学),2014.

[6] Chen L. Apple's early deciduous disease biological control key technology research and its application field(in Chinese)[D]. Xi'an: Northwest University(西安:西北大学),2009.

[7] Shi C X, Huang L, Hui H H, *et al*. Controlling effects of several fungicides against apple leaf spot(in Chinese)[J]. Acta Agriculturae Boreali-Occidentalis Sinica(西北农业学报),2011,20(11):188-191,206.

[8] Jing X K, Lv S, Liu X Y, *et al*. Mapping quantitative trait loci associated with Alternaria leaf blotch susceptibility in a *Malus asiatica* × *M. domestica* interspecific population(in Chinese)[J]. Journal of China Agricultural University (中国农业大学学报),2014, 19(6): 140-147.

[9] Qi N, Wan Y Z, Gao H,*et al*. Development of a SCAR marker linked to apple spot resistance gene from a RAPD marker(in Chinese)[J]. Acta Agriculturae Boreali-Occidentalis Sinica(西北农业学报),2010,19(6): 106-109.

[10] Lv S, Wang Y, Zhang X Z,*et al*. Evaluation of resistance of Malus germplasms to apple alternaria blotch(*Alternaria mali*) and analysis of inheritance for resistance(in Chinese) [J]. Journal of China Agricultural University(中国农业大学学报),2012,17(4): 68-74.

枣疯病植原体对叶片组织发育影响的研究

李晴，刘栋，杨宝江，陈招荣*

（天津农学院园艺园林学院，天津 300384）

摘　要：本研究利用石蜡切片法比较分析感枣疯病部位组织结构与健康组织的差异，结果表明枣疯病植原体侵染导致叶片结构发育受到抑制，叶片厚度、角质层和表皮细胞变薄，薄壁组织细胞显著减少，叶脉组织排列紊乱，形成层和薄壁细胞明显变形和减少，厚角细胞和晶体分泌细胞明显减少。

关键词：枣疯病；植原体；石蜡切片；组织结构

Study on distribution of jujube witches' broom phytoplasma in plant and leaves microstructure of pathogenetic Chinese jujube in spring

Qing Li, Dong Liu, Baojiang Yang, Zhaorong Chen*

(College of Horticulture and Landscape, Tianjin Agricultural University, Tianjin 300384, China)

Abstract: The organization structure of witches' broom and healthy leaves was analyzed by paraffin section method. The results showed that blade structure development has been inhibited in the witches' broom. The thickness of the leaves, cuticles and epidermal cells were significantly reduced. The parenchyma cells were decreased. The arrangement of leaf tissue was disordered. The cambial and epidermal cells were significantly reduced. Collenchyma cells and secretory cell were significantly decreased.

Key words: jujube witches' broom; phytoplasma; paraffin section; organization structure

枣（*Ziziphus jujuba* Mill.）原产于我国黄河流域，栽培历史悠久，是我国特有干果和经济林树种[1]。由于枣树具有抗寒、抗旱、耐涝、耐盐碱等特点及其较高的经济价值，近年来在天津地区种植面积不断扩大，成为发展最快的经济林树种，天津静海县种植枣树 20 多万亩，且呈每年平稳增长趋势，被誉为中国金丝小枣之乡[2]。在枣生产过程中，由植原体（phytoplasma）引起的一类系统性病害——枣疯病（jujube witches' broom）具有毁灭性，遍布于我

基金项目：天津市高等学校科技发展基金计划项目（2140622）；2014 年国家级高等学校大学生创新创业训练计划（201410061006）。

* **通讯作者**：陈招荣，博士，讲师，主要从事植物病毒及植物菌原体虫传机理研究，Tel：022-23791301，E-mail：chenzr@tjau.edu.cn。

国的各大枣区，严重阻碍了枣树产业的发展。

枣疯病的主要症状为花器返祖和枝芽不正常萌发生长，一年多次连续萌生细小枝叶，形成丛枝，叶色黄绿，冬季不落，植株发病初期只在局部枝条上显症，然后扩散到全株，无法正常结果[3]。通过电子显微镜观察发病组织超薄切片可以发现在寄主韧皮部筛管细胞中可以观察到无细胞壁的植原体粒子存在[3]。DAPI 染色技术及显微观察可以发现植原体主要聚集在韧皮部组织当中[3,4]。在植物表现一系列病症的同时或之前，必然先将引起生理上的病变，植原体的侵染影响了植物体的基因表达[5,6]，干扰了植物的正常生理代谢，降低植株叶片叶绿素含量，影响光合作用[6-8]，使各种蛋白表达[6]、激素[9]、酶类[10-12]及其他活性物质和次生代谢物质[12]都发生改变。一般来说病害产生首先发生生理上的病变，然后引起组织结构病变，组织结构病变导致形态上的病变。有大量的研究针对枣疯病的生理病变和形态病变开展，对枣疯病植原体侵染和发病机理提供了大量的证据，但是关于组织结构病变的研究较少。本文以春季萌芽的枝条作为研究对象，利用石蜡切片法观察发病组织和健康组织中叶柄、叶脉以及叶片结构差异，分析枣疯病植原体症状表现及生理生化变化相关的植物组织结构变化。

1 试验材料和方法

1.1 试验材料

春季新生叶及叶柄采集与天津农学院院内及院外绿化带内种植枣树，院内枣树表现为局部丛枝，根据刘孟军等[4]提出的枣疯病分级标准，将其疯枝病情定为Ⅱ。每棵树分别采样，设 3 次重复。

1.2 试验方法

1.2.1 枣疯病植原体 PCR 检测

从不同枣树上采集相同部位完全展开叶片，样品总 DNA 提取参照天根公司植物基因组 DNA 提取试剂盒说明书(DP305-02)。植原体 16S rRNA 基因扩增采用 Lee 等[14]报道的通用引物 R16mF2 \ R16mR1 (5′-CTAGCAAGTCAAGTCGAACGGA-3′, 5′-CTTAACCCCAATCATCGAC- 3′；退火温度 60℃)，反应体系为 20 μL，PCR 产物于 4℃保存备用。nested-PCR 扩增根据 Gunderson[15]等的方法合成 nested-PCR 引物，引物对分别为 R16F2n(5′-GAAACGACTGCTAAGACTGG-3′)和 R16R2n(5′-TGACGGGCGGTGTGTACAAACCCCG-3′)。将前一步获得的 PCR 产物稀释 30 倍后分别作为 DNA 模板，进行 nested-PCR 反应，退火温度为 57℃。取 5 μL PCR 产物进行 1%的琼脂糖凝胶电泳，经 EB 染色后紫外检测仪 254 nm 波长下观察拍照并记录数据。

1.2.2 石蜡切片及分析

5 月 30 日上午 9:00～10:00 从不同枣树上采集相同部位完全展开叶片，分别制作叶片及叶柄石蜡切片，具体流程如下：FAA (38%甲醛、冰醋酸和 70%酒精按 1∶1∶18 体积比混合)固定 24 h 以上。经洗涤、酒精脱水、二甲苯透明、浸蜡、包埋、切片、贴片、二甲苯脱蜡、酒精复水、1%番红 12 h 和 1%固绿 30 s 染色、洗涤和脱水、中性树胶封片干燥后，Leica_DM4000B 显微观察和用 Leica DFC450 CCD 相机拍照。显微照片用图像分析软件 LAS V4.2 处理，测量并计算出角质层、上表皮、下表皮、栅栏组织和海绵组织厚，叶厚等指标，比较叶片主叶脉组织结构。每次取样每个处理制作 5 片以上封片，经镜检拍照后的每个显微图片各指标测量或计算值不少于 12 次，再进行统计分析。

2 结果与分析

2.1 发病植株枣疯病植原体 PCR 检测

采用植原体通用引物对显枣疯症状的植株样品总 DNA 提取液进行 PCR 检测，结果获得 1.4 kb 大小条带，阴性对照与空白对照无相对条带出现。PCR 产物纯化回收稀释 50 倍后，利用 nested-PCR 引物进行扩增。扩增出大小约为 1.2 kb 条带。说明所采集样品中含有植原体。

2.2 枣疯病植原体侵染对植株叶部组织结构的影响

2.2.1 枣疯病植原体侵染对叶片组织结构的影响

健康叶片与发病小叶组织结构比较结果见表 1 和图 1。枣属于耐旱植物，其角质层(A)较厚，健康叶片角质层为(6.91±0.59)μm，而被植原体侵染后角质层(a)厚度显著降低，只为(1.93±1.02)μm，健康叶片枣的叶片栅栏组织(C)发达，海绵组织(D)少且不典型，具有海绵组织向栅栏组织过渡的特征，一般由 5～6 层细胞组成，栅栏组织厚度为(49.47±0.92)μm，海绵组织厚度为(65.40±2.09)μm，发病疯枝小叶叶肉组织细胞明显减少为 4～5 层，且第二层以下细胞排列松散，由长柱形变化为近球形松散海绵组织细胞(d)，细胞间隙明显增大，栅栏组织(c)厚度为(47.09±1.22)μm，海绵组织厚度(43.90±4.22)μm，栅栏组织和海绵组织厚度与健康叶片相比差异显著。植原体侵染后小叶的上表皮(b)和下表皮(e)厚度分别为(17.60±1.00)μm 和(7.81±1.110)μm，与健康叶片(B、E)(20.58± 1.19)μm 和(11.83±0.98)μm 相比也显著减小。整体上发病小叶的叶片厚度明显小于健康叶片，叶片细胞发育受到抑制。在健康叶片的叶肉细胞及近维管束组织周围存在着许多晶体和分泌细胞(F)，但是发病小叶中的这些细胞(f)明显减少。

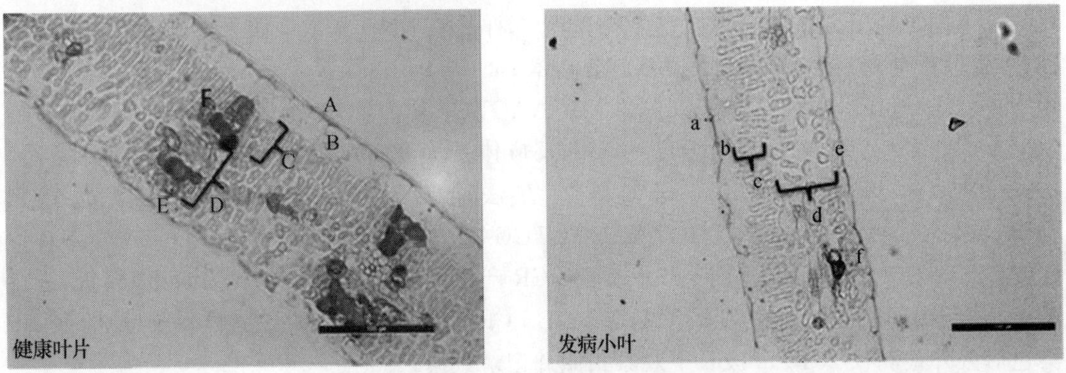

图 1 健康叶片与发病小叶组织结构比较

图中 A，a 为角质层；B，b 为上表皮；C，c 为栅栏组织；D，d 为海绵组织；E，e 为下表皮；F，f 为分泌细胞。标尺为 100 μm。

表 1 健康叶片与病枝条小叶叶片组织结构比较

μm

	栅栏组织	海绵组织	上表皮	下表皮	角质层	叶片厚度
健康叶片	49.47±0.92a	65.40±2.09a	20.58±1.19a	11.83±0.98a	6.91±0.59a	148.91±4.03a
病枝小叶	47.09±1.22 b	43.9±4.22 b	17.60±1.00 b	7.81±1.11 b	1.93±1.02 b	130.36±3.05 b

2.2.2 枣疯病植原体侵染对叶脉结构的影响

枣疯病植原体主要存在于寄主植物维管束组织中，所以本研究比较分析了植原体侵染前后枣叶脉的结构变化情况，比较结果见图 2。枣为三出脉，主脉由维管束和厚角组织组成，木质部在近轴面，韧皮部在远轴面，为外韧维管束。由图可以看出，健康叶片叶脉各部分排列紧密整齐，木质部导管(B)依次发散型排列，紧接着是 2～3 层扁平细胞组成的形成层(C)，韧皮部细胞(D)排列其后，然后是由 4～5 层薄壁细胞(E)和 5～6 层厚角细胞(G)组成的维管束鞘，在薄壁细胞和韧皮部细胞中存在着大量体积大染色深的晶体和分泌细胞(F)。而发病小叶中各结构排列较为松散，其中形成层(c)细胞变形，与韧皮部细胞(d)无明显界线，薄壁细胞(e)严重变形大小不均，厚角细胞(g)明显减少，平均只有 3～5 层。晶体和分泌细胞也明显减少，这与叶片比较结果一致。

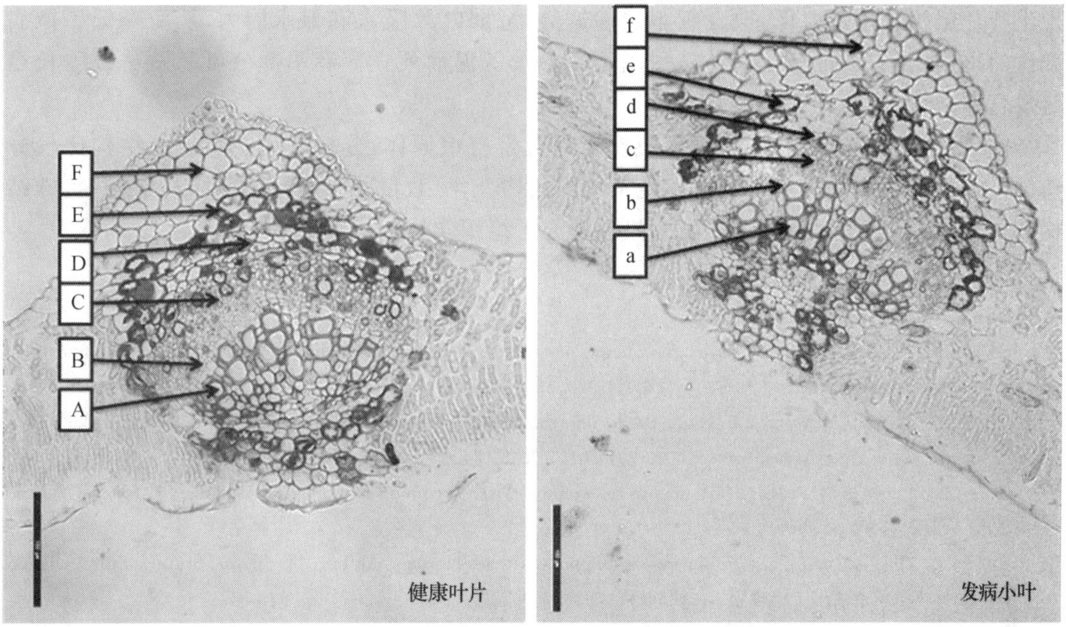

图 2 健康叶片叶脉与发病小叶叶脉组织结构比较

图中 A，a 为导管细胞；B，b 为形成层；C，c 为韧皮部细胞；
D，d 为薄壁细胞；E，e 为晶体和分泌细胞；F，f 为厚角细胞。标尺为 100 μm。

3 讨论与结论

植原体为专性寄生物仅寄生于韧皮部有功能的筛管中，由于受到活动范围或植物抗病性的限制会呈现出长时间局部分布在植株的某些部位，通过病原检测和症状观察人们发现枣疯病症状与病原的存在具有高度相关性[16, 17]。枣树感染枣疯病后通常枝条节间变短、枝叶丛生、叶片黄化并且变小，研究表明患枣疯病病树的净光合速率远远低于正常树水平，叶绿素含量明显降低，光合能力明显下降，光合作用受到明显抑制[7, 8]。病叶的这些异常现象生理病变现象会导致组织病变。为了进一步分析枣疯病植原体对寄主植物症状产生机理，本研究分析了植原体侵染对寄主叶片结构影响，研究发现发病小叶角质层、栅栏组织、海绵组织厚度、叶片厚度以及上表皮厚度与健康叶片相比都明显减小，特别是叶肉细胞由 5～6

层细胞减少到4～5层，海绵组织细胞间隙明显变大，细胞形态有长柱形变为近球形。罹患泡桐丛枝病的病枝小叶组织结构也出现类似的变化。说明叶片组织发育受到抑制，导致光合作用效率变低，物质合成减缓，呼吸作用和蒸腾作用增强，最终出现黄化小叶等症状。植原体主要存在于寄主韧皮部细胞，而且依赖筛管细胞进行系统移动[4]，本研究发现植原体侵染会导致叶脉组织发生紊乱，木质部和韧皮部间的形成层细胞以及韧皮部细胞变形排列杂乱，说明植原体侵染影响了导管和筛管等细胞的分化和形成，最终结果将造成水分的“供给”与蒸腾“消耗”平衡失调，最终导致罹病枝条枯死。薄壁细胞是植物体贮藏物质的主要结构，植原体侵染导致叶脉和叶肉细胞中的薄壁细胞均发生变形和减少，必然会导致组织内淀粉、蛋白以及各种酶类物质减少。在叶肉细胞和维管束组织中发现大量晶体和分泌细胞，由于这些细胞与叶片保水率及硬度密切相关，植原体侵染导致这些细胞大量减少，从症状上就表现出叶片的机械性能降低，改变细胞的渗透势，减弱叶片吸水和保水能力，从而降低了罹病部位的抗性，导致大部分发病丛枝耐受不了冬季低温而死亡。春季部分可萌发丛枝生长迟缓发育不良。

本研究证明了春季发病丛枝可萌发新芽且带有植原体，分析发病叶片组织结构比较分析结果揭示了枣疯病发病后叶片发生了细胞结构变化、数量减少、排列紊乱以及某些特殊结构缺失等一系列变化，为枣疯病发病机理补充了组织学证据。

参考文献

[1] Chen Y J, Wang Q H, Chen C L, *et al*. The present state and history of jujube breeding in China [J]. Journal of Henan Agricultural Sciences(河南农业科学), 1990,12:17-19.

[2] Fan H X, Zhang J H, Wang B B, *et al*. Production status and development trend of Chinese jujube in Jinghai county of Tianjin[J]. Beijing Agriculture(北京农业), 2010, 36:31-34.

[3] Zhou J Y, Liu M Z, Hou B L. Advances in research on Witch-broom disease of Chinese jujube[J]. Journal of Fruit Science(果树科学),1998(4): 354-359.

[4] Liu M Z, Zhao J, Zhou J Y. Grading system of jujube witches'broom symptom [J]. Journal of Agricultural University of Hebei(河北农业大学学报), 2006(1): 31-33.

[5] Fan X, Qiao Y, Han Y, *et al*. High-throughput analysis and characterization of Ziziphus jujube transcriptome jujube witches' broom phytoplasma infected[J]. Phytopathogenic Mollicutes, 2015, 5(S1): S9-S10.

[6] Liu Z, Wang Y, Xiao J, *et al*. Identification of genes associated with phytoplasma resistance through suppressive subtraction hybridization in Chinese jujube[J]. Physiological and Molecular Plant Pathology, 2014, 86: 43-48.

[7] Liu Y Q, Xie N N, Zhao J, *et al*. effects of jujube witches' broom phytoplasma on the chlorophyll content in jujube trees [J]. Plant Protection(植物保护), 2012(3):18-22.

[8] Zhan L J, Feng D Q, Wang Y S, *et al*. Preliminary study on the photosynthetic characteristics of jujube withches' broom trees [J]. Journal of Shanxi Agricultural University(Natural Science Edition)(山西农业大学学报自然科学版), 2010(2): 129-132.

[9] Zhao J, Liu M J, Dai L, *et al*. The variations of endogenous hormones in Chinese jujube infected with witches' broom disease [J]. Scientia Agricultura Sinica(中国农业科学), 2006, 39(11): 2255-2260.

[10] Zhang S H, Gao B J, Wen X J. Studies on the changes of Peroxidase and Phenylalanine ammonia lyase in jujube infected by Phytoplasmas[J]. Plant Protection(植物保护), 2004(5): 59-62.

[11] Niu Q L, Feng D Q, Wang Y S, *et al*. Effects on peroxidase and phenylalanine ammonia-lyase of witches broom of jujube branches under salt-alkali stress [J]. Chinese Agricultural Science Bulletin(中国农学通报), 2010(2): 100-104.

[12] Zhao J H, Wang Y S, Feng D Q, *et al*. Effects of medicament treatment on peroxidase and phenylalanine ammonia-

lyase of jujube branches infected with witches broom disease [J]. Journal of Shandong Agricultural University (Natural Science Edition)(山东农业大学学报:自然科学版), 2010(3): 360-364.
[13] Tian G Z, Zhang X J, Xiong G Y, *et al*. Correlation of callose accumulation in the sieve tubes of paulownia phloem with resistance against witches' broom agent(MLO) [J]. Acta Phytopathologica Sinica(植物病理学报), 1994, 24 (4): 352.
[14] Lee I, Hammond R, Davis R, *et al*. Universal amplification and analysis of pathogen 16S rDNA for classification and identification of mycoplasmalike organisms[J]. Phytopathology, 1993, 83: 834-842.
[15] Gunderson D, Lee I. Ultrasensitive detection of phytoplasmas by nested-PCR assays using two universal primer-pairs[J]. Phytopath Medit., 1996, 35(3): 144-151.
[16] Zhao J, Liu M J, Zhou J Y, *et al*. Distribution and year-round concentration variation of jujube witches' broom (JWB) phytoplasma in the plant of Chinese jujube [J]. Scientia Silvae Sinicae(林业科学), 2006(8):144-146,149.
[17] Shi X Y, Hao S D, Wang H, *et al*. Studies on migration characters of jujube witches' broom phytoplasma into newborn tissues of jujube after overwinter[J]. Journal of Beijing University of Agriculture(北京农学院学报), 2014(3): 56-60.

A transposable element-derived gene *TaS*410 at *Fhb*1 region was associated with Fusarium head blight susceptibility in wheat

Tao Li[1]*, Lei Li[1], Fei Zheng[1], Aili Li[2], Guihua Bai[3,4], Aiai Li[1], Di Wu[1], Long Mao[2]*

([1]Jiangsu Provincial Key Laboratory of Crop Genetics and Physiology/Co-Innovation Center for Modern Production Technology of Grain Crops; Key Laboratory of Plant Functional Genomics of Ministry of Education; Wheat Research Center, Yangzhou University, Yangzhou 225009, China;

[2]National Key Facility for Crop Gene Resources and Genetic Improvement, Institute of Crop Science, Chinese Academy of Agricultural Sciences, Beijing, 100081 China;

[3]Department of Agronomy, Kansas State University, Manhattan, KS 66506, USA;

[4]USDA-ARS Hard Winter Wheat Genetics Research Unit, Manhattan, KS 66506, USA)

Abstract: Fusarium head blight (FHB), incited by *Fusarium graminearum* (*Fg*), causes significant yield losses and quality deterioration in wheat. Due to the complexity of wheat genome and complicated nature of *Fg*-host interactions, the wheat genes regulating interactions with FHB responses remain largely unknown. *Fhb*1 is the most important QTL associated with FHB resistance in wheat. In the current paper, sequencing of RNA from NIL75 (FHB resistant) and NIL98 (FHB susceptible), two near isogenic lines (NILs) contrasting in *Fhb*1 alleles, identified one transcript that was differentially expressed only in NIL98, not in NIL75, and this transcript belongs to the unigene 29451 (temporarily designated as *TaS*410) that locates within *Fhb*1 region between markers *Xsts*189 and *Xsts*142 on ctg0954b of reference sequences and encodes a protein with a domain of unknown function (DUF659) and an hAT (activator superfamily) dimerisation domain. Phylogenetic analysis suggests that *TaS*410 in wheat had the closest phylogenetic relationship with the ortholog in sorghum, and may horizontally transferred from fungus during microbe-plant interactions. The expression of *TaS*410 in NIL98 was spatially and temporally heterogenous , and was upregulated after *Fg* challenge due to the presences of cis-elements responsive to pathogen challenge at the promoter region of *TaS*410. The expression of *TaS*410 is also detected in well-known FHB susceptible varieties including Wheaton, Jagger, Bobwhite, and Clark, but is silenced in a panel of *Fhb*1-containing varieties including Ning7840, Wangshuibai, Huangfangzhu, Baisanyuehuang and Taiwan wheat. In NIL75, a nucleotide deletion on exon 4 of *TaS*410 resulted in a prematured termination of the transcription and gene silencing. These results imply that expression of *TaS*410 results in increased

* Corresponding authors: Tao Li, E-mail: taoli@yzu.edu.cn; Long Mao, E-mail: maolong@caas.cn.

FHB susceptibility whereas silence of this gene may lead to increased resistance in wheat. *TaS*410-specific marker developed here could be used to assist in removal of the susceptible gene/allele to improve FHB resistance in wheat breeding.
Key words: wheat; FHB susceptible gene; *Fhb*1 locus; RNA-seq

Fusarium graminearum FgPLC1 regulates growth, development, stress response, and pathogenicity

Ling Sun, Qili Zhu, Jiajie Lian, Houjuan Xu, Jin Lin, Yuancun Liang*
(College of Plant Protection, Shandong Agricultural University, Taian 271018, China)

Abstract: Phospholipase C (PLC) is an important phospholipid hydrolase that plays critical roles in various cellular processes in eukaryotic cells. To date, little is known about the functions of PLC in morphogenesis and pathogenesis in *Fusarium graminearum*. We constructed deletion mutants for all six *FgPLC* genes in this study. The *FgPLC*1 deletion mutant ($\Delta plc1$) generated by gene replacement showed significant defects in hyphal growth, conidiation, sexual reproduction, pathogenicity, and responses to abiotic stresses. Furthermore, complementation of the $\Delta plc1$ mutant with the full-length *FgPLC*1 gene restored all the defects of the mutant. In addition, analysis of gene expression profile showed that deletion of the *FgPLC*1 gene altered the expression patterns of 1619 genes compared with the wild type strain. In contrast to the $\Delta plc1$ mutant, the deletion mutants of other five genes did not show aberrant phenotypic features including colony morphology and conidiation. Taken together, our results revealed the functions of *FgPLC*1 in *F. graminearum* biology and the potential applications of these findings in development of an effective FHB management protocol.

Key words: *Fusarium graminearum*; *FgPLC*1; growth; pathogenicity

* Ling Sun and Qili Zhu contributed equally to this work. Corresponding author: Yuancun Liang, E-mail: liangyc@sdau.edu.cn.

Fusarium head blight resistance loci in a stratified population of wheat landraces and varieties

Tao Li[1*], Dadong Zhang[2], Xiali Zhou[1,3], Guihua Bai[2,4], Lei Li[1], Shiliang Gu[1]

([1]Jiangsu Provincial Key Laboratory of Crop Genetics and Physiology/Co-Innovation Center for Modern Production Technology of Grain Crops; Key Laboratory of Plant Functional Genomics of Ministry of Education; Wheat Research Center, Yangzhou University, Yangzhou 225009, China; [2]Department of Agronomy, Kansas State University, Manhattan, KS 66506, USA; [3]Zhumadian Academy of Agricultural Sciences, Zhumadian 466000, China; [4]USDA-ARS Hard Winter Wheat Genetics Research Unit, Manhattan, KS 66506, USA)

Abstract: To determine if Chinese and Japanese wheat landraces and varieties have unique sources of Fusarium head blight (FHB) resistance, an association mapping panel of 195 wheat accessions including both commercial varieties and landraces was genotyped with 364 genome-wide simple sequence repeat (SSR) and sequence-tagged site (STS) markers, and evaluated for type Ⅱ FHB resistance in three greenhouse experiments using single floret inoculation. Population structure analysis stratified this population into five groups with Chinese landraces in four groups. Thirty-two of 51 Chinese landraces and 24 of 27 Japanese accessions were placed in one group. Association analysis using a mixed model identified 11 markers having significant associations with FHB resistance in at least two experiments. Most of these markers coincided with known quantitative trait loci (QTL) for FHB resistance with one potentially novel QTL associated with *Xgdm*138-5*DS* and *Xgwm*358-5*DS*. *Xbarc*19-3*AS* was significant in all three experiments, and the frequency of favorable alleles was more than 53%. Chinese landraces and Japanese accessions had more favorable alleles at the majority of reproducible marker loci. Nine QTL combinations were identified according to the number of favorable alleles. Mean FHB severities increased with decreasing numbers of favorable alleles at reproducible loci. The resistance loci characterized here will further diversify the wheat FHB resistance gene pool, and provide breeders with additional sources of resistance for improvement of FHB resistance in wheat.

Key words: allelic variation; association analysis; *Fusarium graminearum*; *Triticum aestivum*; wheat scab

* Corresponding author: Tao Li, E-mail: taoli@yzu.edu.cn.

hph gene inserting into the geome of *Fusarium* effects the biosynthesis of the DON

Cai Yiqiang, Zheng Zhitian, Liu Xiumei, Li Bin, Zhou Mingguo*

(College of Plant Protection, Nanjing Agricultural University, Key Laboratory of Pesticide, Jiangsu Province, Nanjing 210095, China)

Abstract: With the development of molecular biology, transgenic technology is increasingly widespread application in agriculture. People integrate foreign genes into the genome of crops in order to obtain excellent heritable traits. They pay more attention to the safety of the foreign genes expression in host cells for humans, but the influence of cell metabolism of the host is less research when the foreign genes are inserted. In this study, we choosed Hygromycin B phosphotransferase gene (*hph*) which was widely used in molecular biology as the foreign gene inserting into the genome of *Fusarium graminearum* that it could produce secondary metabolites of DON toxin . Then we detected the *Tri*5 gene expression which had a positive exponential relationship with DON production ability by real-time quantitative PCR (RQ-PCR) to evaluate the DON production ability of the transformations. we found that 28% of the 200 transformations we got were down-regulate expressed more than two times and 31% were up-regulate expressed more than two times. Therefore, although the foreign gene itself expression was not harmful to humans, it would trigger the metabolic activity of the host strains, such as secretion of harmful metabolic substances which affected human health.

Key words: *Fusarium graminearum*; foreign genes; DON; real-time quantitative PCR

Identification and characterization of *Phytopythium helicoids* causing stem rot of 'Shatangju' mandarin seedlings in China

Xiaoren Chen[1*], Beibei Liu[1], Yuping Xing[1], Baoping Cheng[2], Yunhui Tong[1], Jingyou Xu[1]

([1]College of Horticulture and Plant Protection, Yangzhou University, Yangzhou 225009, China;
[2]Plant Protection Research Institute, Guangdong Academy of Agricultural Sciences, Guangzhou 510640, China)

Abstract: Citrus mandarin (*Citrus reticulata*) cultivar 'Shatangju' possesses great economic importance in southern China because of its delicious fruits and medicinal peels. During surveys conducted in 2012 and 2013, a stem rot disease were observed on grafted 'Shatangju' seedlings in 21 nurseries located in three regions (Yangjiang City, Sihui City and Deqing County) of Guangdong Province, China. The aim of this study was to characterize the symptoms of seedling stem rot; evaluate the incidence and loss; identify the causal agent and evaluate its pathogenicity and virulence. The disease was present in all surveyed nurseries but incidence varied with location. A *Pythium*-like oomycete species was consistently obtained from symptomatic plants collected in the nurseries surveyed. On the basis of morphological and cultural characteristics and combined phylogenetic analysis of four molecular barcodes including the internal transcribed spacer region (ITS), β-tubulin, mitochondrial *CoxI* and *CoxII* genes, 20 isolates obtained were identified as belonging to the species *Phytopythium helicoids*. Two representative isolates were pathogenic to the scions but rootstocks of 'Shatangju' mandarin seedlings cultivated under nursery conditions and in a greenhouse and symptoms identical to that observed in the field were reproduced. The pathogen was reisolated consistently from the inoculated plants, confirming Koch's postulates. Seedling stem rot and death caused by *P. helicoids* represents a serious threat to 'Shatangju' mandarin nurseries productivity and longevity in southern China.

Key words: 'Shatangju' mandarin; stem rot; *Phytopythium helicoids*; molecular barcode; pathogenicity

* 通讯作者:陈孝仁,男,副教授,主要从事植物真菌病害研究,Tel:0514-87979249,E-mail: xrchen@yzu.edu.cn。

Overexpression of *OsOSM*1 gene enhances rice resistance to sheath blight caused by *Rhizoctonia solani*

Xiang Xue, Zixiang Cao, Xuting Zhang, Yu Wang, YaFang Zhang, Zongxiang Chen, Xuebiao Pan, Shimin Zuo*

(Jiangsu Key Laboratory of Crop Genetics and Physiology/Co-Innovation Center for Modern Production Technology of Grain Crops, Key Laboratory of Plant Functional Genomics of the Ministry of Education, Yangzhou University, Yangzhou 225009, China)

Abstract: Sheath blight (SB), caused by *Rhizoctonia solani*, is one of the most destructive rice diseases worldwide. It has been difficult to generate SB resistant variety through conventional breeding because of typically quantitative nature of rice resistance to SB. Previously, we found that the gene Os12g0569500 (named *OsOSM*1), belongs to TLP-PA subfamily, was strongly induced expression by *R. solani* in SB resistant variety YSBR1. In the study, we identified five genes including *OsOSM*1 that showed high identity on amino acid sequence. Interestingly, only *OsOSM*1 was found highly expressed at rice booting stage and in leaf sheath, which was correlated with SB serious development in rice. Transcription of *OsOSM*1 was strongly induced by methyl jasmonate (MeJA) and reduced by JA biosynthesis inhibitor salicylhydroxamic acid (SHAM), while was not apparently affected by ethephon (ET), salicylic acid (SA) and kinetin (KT). OsOSM1 was localized to plasma membrane. Overexpression of *OsOSM*1 could significantly up-regulate the expression of pathogenesis-related genes and strongly enhanced rice resistance to SB. Remarkably, we did not found apparently alternation of *OsOSM*1 overexpression lines compared with the wild type on morphologies and grain yield. Taken together, our results demonstrate that *OsOSM*1 plays an important role in rice SB resistance and can be used in rice breeding toward SB resistance.

Key words: Rice; *OsOSM*1; gene expression; sheath blight resistance; agronomic traits

* Corresponding author: Shimin Zuo, E-mail: smzuo@yzu.edu.cn.

Overexpression of *OsPGIP1* enhances rice resistance to sheath blight

Xijun Chen[1,2], Yu Chen[2], Bin Xu[1,2], Linan Zhang[2], Zongxiang Chen[1], Yunhui Tong[2], Shimin Zuo[1*], Jingyou Xu[2*]

([1]Jiangsu Key Laboratory of Crop Genetics and Physiology/Co-Innovation Center for Modern Production Technology of Grain Crops, Key Laboratory of Plant Functional Genomics of the Ministry of Education, Yangzhou University, Yangzhou 225009, China;
[2]Horticulture and Plant Protection College, Yangzhou University, Yangzhou 225009, China)

Abstract: Rice sheath blight (SB), caused by necrotrophic pathogen *Rhizoctonia solani*, is one of the most destructive rice diseases and no major resistance genes are available. Polygalacturonase-inhibiting proteins (PGIPs) are extracellular leucine-rich repeat (LRR) proteins and play important roles in plant defense against different pathogenic fungi by counteracting secreted fungal PGs. However, the role of PGIPs in conferring resistance to rice SB remains to be thoroughly investigated. Here we showed that *OsPGIP1* is capable of inhibiting PGs derived from *R. solani*. Our real-time RT-PCR results indicated that resistant rice cultivars YSBR1 and Jasmine 85 express significantly higher levels of *OsPGIP1* than susceptible cultivar Lemont. Our results also show that *OsPGIP1* is most highly expressed at the late tillering stage in the sheath of YSBR1, coinciding with the critical stage of SB development in field. More importantly, the *OsPGIP1* level is highly elevated by inoculation with *R. solani* in resistant cultivars, but not in susceptible Lemont. Overexpression of *OsPGIP1* significantly increased rice resistance to SB and inhibited tissue degradation caused by *R. solani* secreted PGs. Furthermore, *OsPGIP1* overexpression did not affect rice agronomic traits or yield components. Together, our results not only demonstrate the important role of *OsPGIP1* in combatting the rice SB disease, but also provide a new avenue to the improvement of rice SB resistance by manipulating an endogenous gene.

Key words: *Oryzae sativa*; polygalacturonase; polygalacturonase-inhibiting protein gene 1; sheath blight resistance; overexpression

* Corresponding authors: Shimin Zuo, E-mail: smzuo@yzu.edu.cn; Jingyou Xu, E-mail: jyxu@yzu.edu.cn.

Proteomic analysis of lysine acetylation in *Magnaporthe oryzae*

Jinguang Huang[1], Jun Yang[2], Minfeng Xue[3], Hengwei Qian[1], Wenxing Liang[1], You-Liang Peng[2]

([1]College of Agronomy and Plant Protection, Qingdao Agricultural University, Qingdao 266109, China; [2]State key Laboratory of Agrobiotechnology, MOA Key Laboratory of Plant Pathology, China Agricultrual University, Beijing 100193, China; [3]Institute of Plant Protection and Soil Science, Hubei Academy of Agricultural Sciences, Wuhan, 430064, China)

Abstract: *Magnaporthe oryzae*, the causal agent of rice blast, has been used as a model for investigating the interaction between phytopathogenic fungi and plants. Lysine acetylation is emerging as a ubiquitous and conserved posttranslational modificationin living cells. While the role of lysine acetylation in regulating primary metabolism is well-established, its function in secondary metabolism remains largely elusive. To gain insight into the nature, extent and biological function of lysine acetylation in *M. oryzae*, we used the combination of affinity enrichment and high-resolution LC-MS/MS analysis to perform large-scale lysine acetylome analysis. Altogether, 1 283 lysine acetylation sites in 738 protein groups were identified. Intensive bioinformatics analysis was then carried out to annotate those lysine acetylated targets, including protein annotation, functional classification, functional enrichment, functional enrichment-based cluster analysis, etc. Based on the results, further studies following the large-scale lysine acetylome analysis are underway.

Key words: *Magnaporthe oryzae*; proteomic; lysine acetylation

Acknowledgments: This work was supported by the Taishan Scholar Construction Foundation of Shandong Province.

Protoplast preparation and regeneration of *Rhizoctonia cerealis*

Dezhen Zhang[1,2], Xiaoxia Chen[1], Xianyue Gao[1], Jinfeng Yu[1*]
([1]Department of Plant Pathology, College of Plant Protection, Shandong Agricultural University, Tai'an 271018, China; [2]Weifang University of Science and Technology, Shouguang 262700, China)

Abstract: *Rhizoctonia cerealis* is the main pathogen of wheat sharp eyespot in our country. In order to obtain high quality protoplasts for the transformation of *Rhizoctonia cerealis*, the effect factors on protoplast preparation of *R. cerealis* were studied. The results showed that the suitable conditions of protoplast preparation were 6 d mycelium, in the mixture enzymes of 15 mg/mL Lywallayme and 10 mg/mL Snailase, digesting at 30℃ for 4 h, the protoplast yield reached 3.0×10^6 cell/mL. The optimal regeneration conditions of *R. cerealis* protoplast were using SuTC as osmotic stabilizer, adopting TB3 medium for protoplast regeneration, inoculating the protoplasts in the monolayer regeneration medium by mixing the protoplasts with the unsolidified medium, protoplast regeneration rate can reach 58.6%. We believe that this study is an important foundation for deeply exploring the molecular genetics foundation of growing development and the pathogenic mechanism of this pathogen.

Key words: *Rhizoctonia cerealis*; protoplast; preparation; regenetation

* Corresponding author: Yu Jinfeng, E-mail: jfyu@sdau.edu.cn.

Regulation of innate immunity to the fungal pathogen *Fusarium oxysporum* by microRNAs in tomato

Shouqiang Ouyang[1*], Gyungsoon Park[1,4], Hagop Atamian[2,5], Cliff S. Han[3], Jason Stajich[1], Isgouhi Kaloshian[2], Katherine A. Borkovich[1]

([1]Department of Plant Pathology and Microbiology, [2]Department of Nematology Institute for Integrative Genome Biology, 900 University Avenue, University of California, Riverside, CA 92521, USA;
[3]Genome Biology Group, Los Alamos National Laboratory, Los Alamos, NM 87545, USA;
[4]Plasma Bioscience Research Center, Kwangwoon Univeristy,Seoul, Korea;
[5]Department of Plant Biology, University of California, Davis, CA 95616, USA)

Abstract: MicroRNAs (miRNAs) play an important role as regulators of growth and development in plants. Several miRNA families target genes encoding disease resistance proteins, such as nucleotide binding site-leucine-rich repeat (NBS-LRR) plant innate immune receptors. The filamentous fungus *Fusarium oxysporum* f. sp. *lycopersici* causes vascular wilt disease in susceptible tomato (*Solanum lycopersicum*) plants. We explored a possible role for miRNAs in tomato defense against *F. oxysporum* using deep sequencing of small RNA libraries. Comparative miRNA profiling of sensitive (Moneymaker) and resistant (Motelle) tomato cultivars infected with *F. oxysporum* revealed several miRNAs that correlated with disease resistance. In particular, slmiR482f and slmiR5300, two members of the miR482/2118 microRNA superfamily, were repressed during infection with *F. oxysporum*. Northern analysis confirmed that four of the predicted targets of these two miRNAs exhibited increased expression in the resistant Motelle cultivar and encode nucleotide-binding site (NBS) domain-containing proteins. Furthermore, the ability of two predicted mRNA targets of slmiR482f (Solyc08g075630 and Solyc08g076000) and two targets of slmiR5300 (Solyc05g008650 and Solyc09g018220) to be regulated by the miRNAs was confirmed by Agrobacterium-mediated transient co-expression in *Nicotiana benthamiana*. The results also revealed evidence for mRNA cleavage and/or translational inhibition mechanisms during regulation of target gene expression by the microRNAs. A TRV-based virus-induced gene silencing approach in the resistant Motelle cultivar revealed a clear role in resistance to *F. oxysporum* for one target of slmiR482f (Solyc08g075630; Fusarium Resistance Protein-1; FRP-1) and two targets of slmiR5300 (Solyc05g008650; FRP-2 and Solyc09g018220; FRP-3). To our knowledge, this is the first demonstration of microRNAs that mediate resistance to *F. oxysporum* in tomato and first evidence of a role for miR5300 in pathogen resistance in any plant system.

* Corresponding author: Shouqiang Ouyang, E-mail: souyang@ucr.edu.

Transcriptome analysis of *Dlm* mutants reveals the potential formation mechanism of lesion mimic in wheat

Lei Li[1], Xuan Shi[1], Fei Zheng[1], Di Wu[1], Ai'ai Li[1], Fayu Sun[1], Changcheng Li[1], Jincai Wu[2], Tao Li[1*]

([1]Jiangsu Provincial Key Laboratory of Crop Genetics and Physiology/Co-Innovation Center for Modern Production Technology of Grain Crops; Key Laboratory of Plant Functional Genomics of Ministry of Education; Wheat Research Center, Yangzhou University, Yangzhou 225009, China; [2]College of Plant Protection, Yangzhou University, Yangzhou 225009, China)

Abstract: Previous investigations demonstrated that Chinese wheat cv. Ning7840 confers a high level of broad-spectrum resistance to rusts and powdery mildew diseases at the adult plant stage mediated by a spontaneously occurring hypersensitive reaction-like phenotype (also called lesion mimic, LM). In this study, two EMS-induced mutants of cv. Ning7840, *Dlm*1 and *Dlm*2, showed the LM deficiency phenotype. These two mutants were susceptible to powdery mildew. Transcription de novo sequencing of wheat flag leaves of *Dlm*1, *Dlm*2 and wild type at heading stage when the LM trait could be observed on the leaves of wild-type were performed to decipher the molecular mechanisms underlying the phenotypes contrasting in the LM. The results suggest that the diterpenoid-related stress response acts as an indirect regulator of the LM phenotype and broad-spectrum resistance in Ning7840. Oxidative phosphorylation, proteasome, and chlorophyll were involved in the suppression of the LM expression in *Dlm* mutants. This provides new insight into understanding the molecular mechanism of LM and broad-spectrum resistance in wheat.

Key words: wheat; Lesion mimic deficiency mutant; broad-spectrum resistance; diterpenoid biosynthesis; oxidative phosphorylation

* Corresponding author: Tao Li, E-mail: taoli@yzu.edu.cn.

Cloning and prokaryotic expression of a xylanase gene of *Valsa mali* var. *mali*

Xiangpeng Shi, Baohua Li, Caixia Wang*

(Key Laboratory of Integrated Crop, Pest Management of Shandong Province,
College of Agronomy and Plant Protection, Qingdao Agricultural University, Qingdao 266109)

Abstract: *Valsa mali* var. *mali* (*Vmm*), a destructive pathogen of apple tree, secretes a series of cell wall-degrading enzymes (CWDEs) in the infection process, in which the xylanase showed the highest activities and most powerful macerating ability to apple tissues. To better understand the role of xylanase in the pathogenicity of *Vmm*, in this study, a gene encoding xylanase, *Xylanase* Ⅱ, was cloned and identified. The full-length cDNA is 969 bp containing a 5′-untranslated region (5′-UTR) of 135 bp and a 3′-untranslated region (3′-UTR) of 153 bp. The open reading frame (ORF) of 681 bp encodes a protein of 226 amino acid residues with a predicted molecular weight of 23. 8 kD and an isoelectric point of 5. 29, and its predicted molecular formula is $C_{1\,052}H_{1\,620}N_{272}O_{352}S_4$. No putative N-glycosylation site (Asn-Xaa-Thr/Ser) was found. It was identified as a family 11 of glycosyl hydrolase by running a Conserved Domain Search at the National Center for Biotechnology Information (NCBI) website. Amino acid sequences showed the presence of conserved residues at the active site, Glu122 and Glu213 may be catalytic residues, and Trp113 and Trp124 may be involved in substrate binding. Prokaryotic expression results showed that efficient expression of pET-*Xylanase* Ⅱ protein could be realized after induced for 10 h at 15℃ with 0. 1 mmol/L IPTG in *Escherichia coil* Rosetta. Solubility analysis showed that the fused protein mainly existed in the soluble fractions. Western blotting confirmed that the molecular weight of the recombinant pET-*Xylanase* Ⅱ was appropriately 42 kDa, consistent with the predicted result. The enzyme activity of the recombinant protein was performed, and high xylanase activates were observed when reaction temperature was 50℃ and reaction time was 30 min. In the present study, *Xylanase* Ⅱ of *Vmm* was successfully identified, and prokaryotic expression protein with high activity was achieved, which provide a basis for better elucidating *Vmm* pathogenicity mechanisms and lay a foundation for developing more effective disease-management strategies.

基金项目:国家自然科学基金(31272001 和 31000891);现代农业产业技术体系建设专项资金(CARS-28);山东省科技攻关计划(2010GNC10918)。

作者简介:史祥鹏(1990—),男,山东潍坊人,硕士研究生,研究方向果树病理学,E-mail:xiangpengshi@126. com。

* **通讯作者**:王彩霞,教授,主要从事果树病害及分子植物病理学研究,E-mail:cxwang@qau. edu. cn。

Cap-independent translation of ORF1 (p35) encoded by *Tobacco bushy top virus*

Deya Wang, Chengming Yu, Guolu Wang, Xuefeng Yuan*
(College of Plant Protection, Shandong Agricultural University, Tai'an 271018, China)

Abstract: *Tobacco bushy top virus* (TBTV), a member of *Umbravirus*, contains a positive single genomic RNA with 4 152 nts. The genome of TBTV lacked 5′ cap and 3′ poly (A) tail and encoded 4 proteins including components of replicase (p35 and p63), and movement proteins (p26 and p27). For the cap-independent translation of p35, previous report showed the existence of a BTE-like element in TBTV by using firefly luciferase (Fluc) reporter vector. We confirmed the BTE-like element in full-length TBTV through in vitro translation in wheat germ extract (WGE). Furthermore, we used forward and reverse genetics methods to identify new elements regulating the expression of p35. The expression of p35 among different TBTV isolates showing severe or mild pathogenicity had 3 fold variations. Consequently, the regions altering the expression of p35 were mapped within 5′ terminal 500 nt and internal 600 nt regions surrounding BTE-like elements by constructing chimeric TBTV. The 3′ proximal 120 nt regions located downstream of BTE-like element were also played role in the translation of p35 by deletion mutagenesis. The 3′ terminal 120 nt regions contained 5 hairpins and three RNA pseudoknots through in-line probing and their roles on the translation of p35 will be further identified. In a word, it is suggested that at least three cis-elements including BTE-like element were associated with the cap-independent translation of p35.

Key words: *Tobacco bushy top virus*; cap-independent translation; BTE-like element; in-line probing

基金项目:国家自然科学基金项目(31370179);山东省自然科学基金项目(ZR2013CM015)。

作者简介:王德亚,博士研究生,Tel:0538-8205608。

* **通讯作者**:原雪峰,教授,研究方向:植物分子病毒学和植物 RNA 病毒的基因表达调控,Tel:0538-8205608,E-mail:snowpeak77@163.com。

Cis-elements of -1 programmed ribosome frameshift responsible for the expression of RdRp in *Tobacco bushy top virus*

Chengming Yu, Deya Wang, Guolu Wang, Xuefeng Yuan*

(College of Plant Protection, Shandong Agricultural University, Tai'an 271018, China)

Abstract: *Tobacco bushy top virus* (TBTV) has a (+) ssRNA genome with 4 152 nucleotides. The viral genome without 5′-cap and 3′-poly (A) encoded four putative open reading frames (ORFs), in which ORF1 (p35) and ORF2 (p63) are components of replicase based on common charactericstic of *umbravirus*. Through in vitro translation of TBTV in wheat germ extract (WGE), it showed that ORF2 (p63) was expressed through -1 type frameshift mechanism and existed as ORF1-fused pattern resulting in the production of p98, which may function as RNA-dependent RNA polymerase (RdRp). Through comparation of sequences, mutagenesis and in vitro translation, we identified the essential heptanucleotide slippery sequences at^{946}GGATTTT, whose site-mutations decreased the frameshift ratio to about 30% of wild type. We also tested the optimal distance between heptanucleotide slippery sequences and potential RNA structure located 6 nt downstream, showing that 9 nt is the upper limit. Furthermore, long distance RNA-RNA interaction was found though electrophoretic mobility shift assay (EMSA). The 3′ terminal 200 nt could form RNA-RNA interaction with potential RNA structure located downstream of heptanucleotide slippery sequences. In-line probing showed that one of the RNA-RNA interaction sites is located at the 3′ proximal hairpin, which is also responsible for replication in similar genus. The local RNA structure and detailed long distance RNA-RNA interaction, which is essential for -1 type frameshift, will be further identified.

Key words: *Tobacco bushy top virus*; ribosomal frameshift; RNA-RNA interaction; in-line probing

---

基金项目:国家自然科学基金项目(31370179);山东省自然科学基金项目(ZR2013CM015)。

作者简介:于成明,博士研究生,Tel:0538-8205608。

* **通讯作者**:原雪峰,教授,研究方向:植物分子病毒学和植物 RNA 病毒的基因表达调控,Tel:0538-8205608,E-mail:snowpeak77@163.com。

Complete nucleotide sequence and genome organization of *Fig fleck-associated virus*-2, a novel member of the family Tymoviridae

Zhen He[1,2*], Mahmut Mijit[2], Zhixiang Zhang[2], Shifang Li[2*]
([1]Department of Plant Pathology, School of Horticulture and Plant Protection, Yangzhou University, Yangzhou 225009, China; [2]State Key Laboratory of Biology of Plant Diseases and Insect Pests, Institute of Plant Protection, Chinese Academy of Agricultural Sciences, Beijing 100193, China)

Abstract: The complete nucleotide sequence of a novel member of the Tymoviridae family of plant viruses with a distinct genome structure was determined and named Fig fleck-associate virus-2 (FFkaV-2) based on the sequencing results. The 6 723 nucleotide viral genome comprises three open reading frames (ORFs) excluding a terminal poly(A) tail. ORF1 encodes a polypeptide of 1 878 amino acid (aa) residues with four conserved replication-associated protein domains. ORF2 encodes a putative movement-associated protein of 461 aa residues. ORF3 encodes a unique two-domain protein with an N-terminal domain of unknown function and a putative C-terminal coat protein. Pairwise diversity identified *Fig fleck-associated virus* (FFkaV) as the closest homolog, and phylogenetic analysis clustered FFkaV-2 into the *Maculavirus* group based on RP and CP aa sequences.

* Corresponding authors: Shifang Li, E-mail: sfli@ippcaas. cn; Zhen He, E-mail: hezhen@yzu. edu. cn.

Developmentally regulated plasma membrane protein of *Nicotiana benthamiana* contributes to potyvirus movement and transports to plasmodesmata via the early secretory pathway and the actomyosin system

Chao Geng[1], Qianqian Cong[1], Xiangdong Li[1*], Anli Mou[1], Rui Gao[1], Jinliang Liu[2], Yanping Tian[1]

([1]Department of Plant Pathology, College of Plant Protection, Shandong Agricultural University, Tai'an 271018, China; [2]College Of Plant Sciences, Jilin University, Changchun 130062, China)

Abstract: Intercellular movement of plant viruses requires both viral and host proteins. Previous studies have demonstrated that the P3N-PIPO and cylindrical inclusion (CI) proteins were required for potyvirus cell-to-cell movement. Here, we provided genetic evidence showing that *Tobacco vein banding mosaic virus* (TVBMV; genus *Potyvirus*) mutant carrying a truncated PIPO domain of 58 aa residues could move between cells and induce systemic infection in *N. benthamiana* plants; mutants carrying a PIPO domain of 7, 20 or 43 aa residues failed to move between cells and cause systemic infection in this host plant. Interestingly, the movement-defective mutants produced progeny that eliminated the previously introduced stop codons and thus restored their systemic movement ability. We also presented evidences showing that a developmentally regulated plasma membrane protein of *Nicotiana benthamiana* (referred to as NbDREPP) interacted with both the P3N-PIPO and CI of the movement-competent TVBMV. The knockdown of *NbDREPP* gene expression in *N. benthamiana* impeded the cell-to-cell movement of TVBMV. NbDREPP was shown to co-localize with TVBMV P3N-PIPO and CI at plasmodesmata (PD) and traffick to PD via the early secretory pathway and the actomyosin motility system. We also showed that myosin XI-2 is specially required for transporting NbDREPP to PD. In conclusion, NbDREPP is a key host protein within the early secretory pathway and the actomyosin motility system that interacts with two movement proteins and influences virus movement.

This work was supported in part by grants from the National Natural Science Foundation of China (30971895 and 31011130031), and Special Research Funds for Doctoral Program from Ministry of Education, China (SRFDP; 20123702110013).

* Corresponding author: Xiang-Dong Li, E-mail: xdongli@sdau.edu.cn.

Dimeric artificial microRNAs mediate highly efficient RSV and RBSDV resistance in transgenic rice plants

Lin Sun, Chao Lin, Yunzhi Song, Jinwen Du, Fujiang Wen*, Changxiang Zhu*

(State Key Laboratory of Crop Biology, Shandong Key Laboratory of Crop Biology, College of Life Sciences, Shandong Agricultural University, Tai'an 271018, China)

Abstract: Infection of *Rice stripe virus* (RSV) and *Rice black streaked dwarf virus* (RBSDV) on rice results in severe economic losses on crop production. As the natural resistance resources against these viruses are limited, it is imperative to elaborate a biotechnological approach that will provide effective and safe immunity to RSV and RBSDV. In this study, we generated three dimeric amiRNA precursor expression vectors (pamiR-M, pamiR-3 and pamiR-U) that simultaneously targeting RSV and RBSDV based on the structure of rice osa-MIR528 precursor. The transgenic rice plants were obtained by *Agrobacterium tumefaciens*-mediated transformation and proved to express amiRNAs successfully. The viral challenge assays revealed that these transgenic plants could be conferred highly specific resistance against RSV and RBSDV infection simultaneously. The amiR-RSVU and the amiR-RBSDVU induced the highest virus resistance against RSV and RBSDV, respectively. The target viral RNAs could be cleavaged by the amiRNAs and there was a relationship between the resistance level and the accumulation of amiRNA expression. A Northern blot assay verified that silencing was induced by the original amiRNAs and could be bilaterally extended by the siRNA pathway. The amiRNA together with the secondary siRNAs mediated the degradation of viral RNAs. Genetic stability assay showed that transgene and amiRNA-mediated virus resistance can be stably inherited toT2.

Key words: artificial microRNA; *Rice stripe virus*; *Rice black streaked dwarf virus*; multiple virus resistances; transgenic rice

This work was partially supported by the National Special Grad Project of the Genetically Modified New Seeds Cultivation (No. 2014ZX08001-002) and the National Natural Science Foundation of China (No. 31272113).

* Corresponding authors: Changxiang Zhu, E-mail: zhchx@sdau.edu.cn; Fujiang wen, E-mail: fjwen@sdau.edu.cn.

Genetic structure of populations of *Sugarcane streak mosaic virus* in China and comparison with isolates from India

Zhen He[1,2*], Ryosuke Yasaka[3,4], Shifang Li[2*], Kazusato Ohshima[3,4]

([1]Department of Plant Pathology, School of Horticulture and Plant Protection, Yangzhou University, Yangzhou 225009, China;

[2]State Key Laboratory of Biology of Plant Diseases and Insect Pests, Institute of Plant Protection, Chinese Academy of Agricultural Sciences, Beijing 100193, China;

[3]Laboratory of Plant Virology, Department of Applied Biological Sciences, Faculty of Agriculture, Saga University, 1-banchi, Honjo-machi, Saga, 840-8502, Japan;

[4]The United Graduate School of Agricultural Sciences, Kagoshima University, 1-21-24, Kagoshima, 890-0065, Japan)

Abstract: *Sugarcane streak mosaic virus* (SCSMV) causes mosaic and streak symptoms on infected plants, and is one of the most important viruses infecting sugarcane and sorghum crops. SCSMV is a member of the genus *Poacevirus* in the family *Potyviridae*. Ten SCSMV isolates were collected from sugarcane plants showing mosaicism and streaking in southern China from 2009—2011. Sequence-based phylogenetic and population genetic analyses were conducted using four partial genomic sequences covering the full genomes. These analyses were used to estimate the subpopulation differentiation and divergence within the Chinese virus population, and were compared with isolates from India. SCSMV-infected sugarcane plants in the field commonly harbor virus quasispecies (mutant cloud), and often have mixed infections with the same virus isolates. Inter- and intra-lineage recombination sites were identified in the P1, HC-Pro, CP, and 3′ non-coding regions of the Chinese isolates. All the Chinese non-recombinant isolates fell into at least nine lineages, and many clustered with Indian isolates. However, estimates of genetic differentiation and gene flow indicated that the SCSMV populations in China and India are genetically independent. Our genetic study of a poacevirus population in South Asia regions indicates the importance of such studies for the science-based design of potential virus control strategies.

* Corresponding authors: Shifang Li, E-mail: sfli@ippcaas.cn; Zhen He, E-mail: hezhen@yzu.edu.cn.

Mapping of the minimal epitopes for three coat protein specific monoclonal antibodies commonly used to detect *Potato virus Y*

Yanping Tian[1,2], Jussi Hepojoki[3], Harri Ranki[3], Hilkka Lankinen[3], Jari P. T. Valkonen[1]

([1]Department of Agricultural Sciences, Plant Pathology Laboratory, University of Helsinki, Helsinki, Finland; [2]Department of Plant Pathology, Shandong Agricultural University, Tai'an 271018, China; [3]Department of Virology, Peptide and Protein Laboratory, Infection Biology Research Program, Haartman Institute, University of Helsinki, Helsinki, Finland)

Abstract: *Potato virus* Y (PVY) (genus *Potyvirus*) is the most economically damaging in Solanaceous plants including potato, tobacco and pepper. Enzyme-linked immunosorbent assay (ELISA) has been widely used for detection of PVY. Mab1128, Mab1129 and Mab1130 were three of widely used commercial monoclonal antibodies to distinguish different PVY strains. In this study we mapped three minimal epitopes within coat protein for those three antibodies using peptide scanning, alanine replacement, and residue deletion. All mapped epitopes located within 30 N-terminal residues of coat protein. Sequences alignments showed that these epitopes are relatively conserved. Mutagenesis results of the most variable position within the epitopes indicated that some of those virus isolates could be escaped the detection.

Reference

Tian Y P, Hepojoki J, Ranki H, Lankinen H, Valkonen J P T. Analysis of *Potato virus* Y coat protein epitopes recognized by three commercial monoclonal antibodies. PLoS ONE, 2014, 9: e115766.

Pokeweed antiviral protein (PAP) increases plant resistance to *Tobacco mosaic virus* infection in *Nicotiana benthamiana*

Feng Zhu*, Yizhong Yang
(College of Horticulture and Plant Protection, Yangzhou University, Yangzhou 225009, China)

Abstract: Ribosome-inactivating proteins (RIPs) are toxic *N*-glycosidases that depurinate the universally conserved α-sarcin loop of large eukaryotic and prokaryotic rRNAs, thereby arresting protein synthesis at translation. RIPs are widely distributed in various plant species and within a variety of different tissues. As for plants, RIPs have been linked to defense by antibacterial, antiviral, antifungal, and insecticidal properties demonstrated in *vitro* and in transgenic plants. Pokeweed antiviral protein (PAP) is a 29 kDa type I RIP isolated from the leaves of the pokeweed plant (*Phytolacca americana*). Our recent study demonstrated that the PAP purified from the leaves of the pokeweed plant has distinct antiviral activity. *Nicotiana benthamiana* leaves pre-treated with 0. 2 mg/mL PAP three days before inoculated with *Tobacco mosaic virus* (TMV) showed less-severe symptom and less reactive oxygen species (ROS) accumulation compared with that inoculated with TMV only but no PAP pre-treatment. Quantitative real-time PCR analysis revealed that the replication levels of TMV were lower in the PAP-treated leaves compared with the control plants at 7 days post inoculation. Out results indicated that PAP increases plant resistance to TMV infection. We think that PAP may act on ribosomes of infected plant cells, thereby inhibit the synthesis of viral protein. The results will be useful to design appropriate strategies for transgenic resistance in crop plants and PAP could possibly be exploited in crop protection.

This work was supported by the National Natural Science Foundation of China (Grant no. 31500209) and Natural Science Foundation of the Higher Education Institutions of Jiangsu Province of China (Grant no. 15KJB210007).

* Corresponding author: Feng Zhu, E-mail: zhufeng@yzu. edu. cn.

Recombination of strain O segments to HCpro-encoding sequence of strain N of *Potato virus Y* modulates necrosis induced in tobacco and in potatoes carrying resistance genes *Ny* or *Nc*

Yanping Tian[1,2], Jari P. T. Valkonen[1]

([1]Department of Agricultural Sciences, Plant Pathology Laboratory, University of Helsinki, Helsinki, Finland; [2]Department of Plant Pathology, Shandong Agricultural University, Tai'an 271018, China)

Abstract: Hypersensitive resistance (HR) to strains O and C of *Potato virus Y* (PVY, genus *Potyvirus*) is conferred by potato genes Ny_{tbr} and Nc_{tbr}, respectively; however, PVY N strains overcome these resistance genes. The viral helper component proteinases (HCpro, 456 amino acids) from PVY^N and PVY^O are distinguished by an eight-amino-acid signature sequence, causing HCpro to fold into alternative conformations. Substitution of only two residues (K269R and R270K) of the eight-amino-acid signature in PVY^N HCpro was needed to convert the three-dimensional (3D) model of PVY^N HCpro to a PVY^O-like conformation and render PVY^N avirulent in the presence of *Nytbr*, whereas four amino acid substitutions were necessary to change PVY^O HCpro to a PVY^N-like conformation. Hence, the HCpro conformation rather than other features ascribed to the sequence were essential for recognition by Ny_{tbr}. The 3D model of PVY^C HCpro closely resembled PVY^O, but differed from PVY^N HCpro. HCpro of all strains was structurally similar to β-catenin. Sixteen PVY^N605-based chimeras were inoculated to potato cv. Pentland Crown (Ny_{tbr}), King Edward (Nc_{tbr}) and Pentland Ivory (Ny_{tbr}/Nc_{tbr}). Eleven chimeras induced necrotic local lesions and caused no systemic infection, and thus differed from both parental viruses that infected King Edward systemically, and from PVY^N 605 that infected Pentland Crown and Pentland Ivory systemically. These 11 chimeras triggered both Ny_{tbr} and Nc_{tbr} and, in addition, six induced veinal necrosis in tobacco. Further, specific amino acid residues were found to have an additive impact on necrosis. These results shed new light on the causes of PVY-related necrotic symptoms in potato.

Reference

Tian Y P, Valkonen J P T. Recombination of strain O segments to HCpro-encoding sequence of strain N of *Potato virus Y* modulates necrosis induced in tobacco and in potatoes carrying resistance genes *Ny* or *Nc*. Mol Plant Pathol., 2015,16: 735-747.

Studies on *Cassava brown streak disease-associated virus*

Deusdedith R. Mbanzibwa[1,2,3], Yanping Tian[1,4], Settumba B. Mukasa[2], Jari P. T. Valkonen

([1] Department of Applied Biology, University of Helsinki, P. O. Box 27, Helsinki FIN-00014, Finland;

[2] Department of Crop Science, Makerere University, P. O. Box 7062, Kampala, Uganda;

[3] Mikocheni Agricultural Research Institute, P. O. Box 6226, Dar es Salaam, Tanzania;

[4] Department of Plant Pathology, Shandong Agricultural University, Tai'an 271018, China)

Abstract: Cassava is the second important root crop in the world after potato. It is a major source of carbohydrates in the diet of millions of people and grown as a famine reserve crop. Cassava brown streak disease (CBSD) caused by *Cassava brown streak virus* (CBSV, Genus *Ipomovirus*) was a big problem in cassava plant in East Africa. It causes not only significant losses in yield but also necrosis in in storage roots make them unsalable. Samples obtained from Lake Victoria basin in Uganda and Tanzania had been considered as different genetic groups of CBSV based on the analysis of their coat protein sequences. The complete genome of CBSV showed that it contains a P1 proteinase that suppresses RNA silencing but HCpro. In addition, a putative Maf/Ham1 pyrophosphatase was found between NIb and coat protein. Comparison of 12 complete genomes of CBSD-associated viruses suggested the genetic variation was large enough for demarcation of them into two distinct species: CBSV and *Ugandan Cassava brown streak virus* (UCBSV). To distinguish these two viruses, two different primer pairs that simultaneously detect UCBSV and CBSV isolates were developed. Our study has contributed the understanding of evolution of CBSV and UCBSV, moreover, the detection tool will be useful in plant breeding.

References

[1] Mbanzibwa D R, Tian Y P, Tugume A K, *et al*. Genetically distinct strains of *Cassava brown streak virus* in the Lake Victoria basin and the Indian ocean coastal area of East Africa. Arch Virol., 2009, 154: 353-359.

[2] Mbanzibwa D R, Tian Y P, Mukasa S B, *et al*. *Cassava brown streak virus* (*Potyviridae*) encodes a putative Maf/Ham1 pyrophosphatase implicated in reduction of mutations and a P1 proteinase that suppresses RNA silencing but contains no HCpro. J. Virol., 2009, 83: 6934-6940.

[3] Mbanzibwa D R, Tian Y P, Tugume A K, *et al*. Simultaneous virus-specific detection of the two *Cassava brown streak-associated viruses* by RT-PCR reveals wide distribution in East Africa, mixed infections, and infections in *Manihot glaziovii*. J. Virol. Methods., 2011, 171: 394-400.

[4] Mbanzibwa D R, Tian Y P, Tugume A K, *et al*. Evolution of *Cassava brown streak disease-associated viruses*. J. Gen. Virol., 2011, 92: 974-987.

Sequence analysis and functional characterization of the antifungal biosynthetic pathway from *Burkholderia pyrrocinia* strain Lyc2

Xiaoqiang Wang [1,2,3*], Aixin Liu[1*], Jin Liu[1,2], Xiaoqing Yu[1], Peng Deng[3], Sonya M. Baird[3], Xiangdong Li[1,2], Shi'en Lu[3]

([1]Department of Plant Pathology, College of Plant Protection, Shandong Agricultural University, Tai'an, China; [2]Collaborative Innovation Centre for Annually High Yield and High Efficiency Production of Wheat and Corn, Shandong Agricultural University, Tai'an, China; [3]Department of Biochemistry, Molecular Biology, Entomology and Plant Pathology, Mississippi State University, Mississippi State, Mississippi, United States of America)

Abstract: Bacterial strain Lyc2 was isolated from tobacco rhizosphere and showed strong antifungal activities. The 16S rRNA and multilocus sequence typing (MLST) analyses revealed that strain Lyc2 belongs to *Burkholderia pyrrocinia*. Bioassay results indicated Lyc2 showed significant antifungal activities against a broad range of plant and animal fungal pathogens. Mutagenesis analysis revealed that a 55.2 kb gene cluster was essential for antifungal activities of strain Lyc2. In this study, the mutant Lyc2-MT11 (orf11::Tn5) showed significant reduction in antifungal activity against *Geotrichum candidum* compared with that of the wild-type strain. Analysis of the protein sequence suggests that the *orf*11 gene encodes a flavin-dependent monooxygenase. HPLC was used to verify the production of antifungal compounds. Multiple antibiotic and secondary metabolized biosynthesis gene clusters predicated by antiSMASH revealed the broad spectrum of antimicrobials activities of the strain. Our results revealed the mechanisms of antifungal activities of strain Lyc2 and expand our knowledge about production of antifungal compounds in the bacteria *Burkholderia pyrrocinia*.

Key words: antifungal activity; *Burkholderia pyrrocinia*; strain Lyc2

* These authors are contributed equally to this work.

First report of corn whorl rot caused by *Serratia marcescens* in China

Xiaoqiang Wang[1,2], Tao Bi[1,2], Xiangdong Li[1,2], Liqun Zhang[3], Shi'en Lu[4]

([1]Department of Plant Pathology, College of Plant Protection, Shandong Agricultural University, Tai'an 271018, China; [2]Collaborative Innovation Center for Annually High Yield and High Efficiency Production of Wheat and Corn, Shandong Agricultural University, Tai'an 271018, China; [3]Department of Plant Pathology, China Agricultural University, Beijing 100193, China; [4]Department of Biochemistry, Molecular Biology Entomology and Plant Pathology, Mississippi State University, Mississippi State, MS 39762, USA)

Abstract: Whorl rot is a novel disease of corn found in the Huang-Huai-Hai Plain, China. Common symptoms of the disease in fields include yellowing and water-soaked brown necrosis of young leaves in the whorl of corn plants, which often results in rot of the whorl. Bacterial streaming was always observed from diseased samples. Bacterial isolates were obtained from symptomatic tissue and further confirmed to be the causal agent of the disease using Koch's Postulates. Sequence analysis of 16S rDNA, housekeeping gene groES-groEL, BIOLOG and API 20NE tests revealed that the isolate B3R3 belongs to the bacterium *Serratia marcescens*. None of the corn cultivars evaluated showed acceptable resistance to the disease. To our knowledge, this is the first report on corn whorl rot caused by *Serratia marcescens*.

Isolation and characterization of a azoxystrobin-degrading bacterial strain *Ochrobactrum anthropi* SH14

Shaohua Chen[1*], Yinyue Deng[1], Fei He[1], Lianhui Zhang[1,2,3]

([1]Guangdong Province Key Laboratory of Microbial Signals and Disease Control, South China Agricultural University, Guangzhou 510642, China;

[2]Institute of Molecular and Cell Biology, Agency for Science, Technology and Research (A * STAR), 61 Biopolis Drive, Proteos, Singapore 138673, Republic of Singapore;

[3]Department of Biological Sciences, National University of Singapore, Republic of Singapore)

Abstract: Continuous use of the strobilurin fungicide azoxystrobin (AZX) has resulted in serious environmental contamination problems. We report here that a novel bacterial strain SH14, which was isolated from wastewater treatment system using an enrichment culture technique, was able to degrade and utilize AZX as the carbon and nitrogen source in minimal medium. Based on the morphology, physio-biochemical characteristics, 16S rDNA gene analysis, and API 20 NE systems, strain SH14 was identified as *Ochrobactrum anthropi*. Strain SH14 degraded 86.3% of 50 mg • L^{-1} AZX within 5 days. The optimum conditions for AZX degradation were determined to be 30.2°C, pH 7.9 and inocula amount 0.2 g dry wt • L^{-1} using response surface methodology. Andrews equation was used to describe the special degradation rate at different initial concentrations. It was observed that strain SH14 degraded AZX up to a concentration of 400 μg • mL^{-1} with a maximum specific degradation rate (q_{max}), half-saturation constant (K_s) and inhibition constant (K_i) of 0.612 2 d^{-1}, 6.829 1 μg • mL^{-1} and 188.468 0 μg • mL^{-1}, respectively. The critical inhibitor concentration was estimated to be 35.875 7 μg • mL^{-1}. Moreover, strain SH14 participated in efficient degradation of a wide range of strobilurin fungicides including kresoxim-methyl, pyraclostrobin, and trifloxystrobin, which similar to AZX are also widely used pesticides with environmental contamination problems with the degradation process following the first-order kinetic model. Our results highlight the promising potentials of strain SH14 in bioremediation of AZX-contaminated environments.

Key words: azoxystrobin; *Ochrobactrum anthropi*; biodegradation; kinetics

* Corresponding author: Shaohua Chen, Tel: +86-20-85288229, Fax: +86-20-85280292, E-mail: shchen@scau.edu.cn.

Tat pathway-mediated translocation pathway is essential for antibacterial activity of *Pseudomonas fluorescens* XW10 against *Ralstonia solanacearum*

Xiaoqiang Wang[1,2], Peng Deng[2], Sonya M. Baird[2], Xiangdong Li[1], Shi'en Lu[2]

([1] Department of Plant Pathology, College of Plant Protection, Shandong Agricultural University, Tai'an, China; [2] Department of Biochemistry, Molecular Biology, Entomology and Plant Pathology, Mississippi State University, Mississippi State, MS, USA)

Abstract: Strain XW10 was isolated from soybean rhizosphere and showed a broad-spectrum of antimicrobial activities, especially against *Ralstonia solanacearum*. Phylogenetic analysis of the 16S *rRNA*, *gyrB*, *rpoB* and *rpoD* genes revealed that strain XW10 belongs to the genus *Pseudomonas* but apparently different from any of established species of *Pseudomonas*, which indicated that XW10 could be a new species of *Pseudomonas*. To characterize the genes dedicated to antibacterial activities against *R. solanacearum*, a Mini Tn5-mutation library of strain XW10 was constructed. The mutant XW10-5206 was eliminated in antibacterial activity against *R. solanacearum* when compared to the wild-type strain XW10. Sequence analysis of the mutant indicated that the gene disrupted by transposon shared high sequence identity (98%) to *TatA* gene in twin-arginine translocase (Tat) secretion system predicated in *Pseudomonas fluorescens* strain Pf0-1 (GenBank accession number: CP000094). Constitutive expression plasmid pUCP26-TatA was constructed and electroporated into the mutant XW10-5206. As expected, introduction of plasmid pUCP26-TatA restored the antibacterial activity of the mutant against *R. solanacearum* to the wild-type level. Thus the Tat pathway-mediated translocation pathway is essential for antibacterial activity of *Pseudomonas* sp. XW10 against *R. solanacearum*.

Key words: Antimicrobial activity; *Pseudomonas*; *Ralstonia solanacearum*; twin-arginine translocase secretion system

Large-scale identification of wheat genes resistant to cereal cyst nematode *Heterodera avenae* using comparative transcriptomic analysis

Ling'an Kong[1], Duqing Wu[1], Wenkun Huang[1], Huan Peng[1], Gaofeng Wang[1], Jiangkuan Cui[1], Shiming Liu[1], Zhigang Li[2], Jun Yang[2], Deliang Peng[1]*

([1] State Key Laboratory for Biology of Plant Diseases and Insect Pests, Institute of Plant Protection, Chinese Academy of Agricultural Sciences, Beijing 100193, China;
[2] State Key Laboratory of Agrobiotechnology and MOA Key Laboratory of Plant Pathology, China Agricultural University, Beijing 100193, China)

Abstract: Cereal cyst nematode *Heterodera avenae*, an important soil-borne pathogen on wheat, causes numerous yield losses annually worldwide, and utilization of resistant cultivars is one of the best strategies to control it. However, target genes were scare available for breeding resistant cultivars. To identify more applicable candidate resistance genes against *H. avenae* for breeding cultivars, RNA-sequencing for comparative transcriptomic analysis was performed by using an incompatible wheat cultivar VP1620 infected with *H. avenae* at early stage (24 h, 3 d and 8 d, respectively), while the cultivar WEN19 was used as compatible controls. Infection assays showed that VP1620 failed to block penetration and migration of *H. avenae* but delay transition of developmental stages, and eventually leading to significant reduction in cyst formation. A novel strategy was then established to generate clean data by removing the transcripts of *H. avenae* within raw data before assembly. Two types of expression profiles were established to predict candidate resistance genes. During the uncoordinated expression profiles with transcript abundance as a standard, totally 424 candidate resistance genes including 302 overlapping genes and 122 VP1620-specific genes were identified. The genes with similar expression patterns were further classified according to the scales of changed transcript abundances, and 182 genes were totally identified as supplemental candidate resistance genes. Functional characterizations revealed that diverse defense-related pathways were responsible for resistance against *H. avenae*. Moreover, phospholipase was found to involve in many defense-related pathways and localize in the connection position. Furthermore, strong bursts of reactive oxygen species (ROS) within VP1620 roots infected with *H. avenae* were induced

---

This project was supported by grants from the Natural Science Foundation of China (31301645) and the 973 project (2013CB127502).

* Corresponding author: Deliang Peng, Tel: 86-10-62815611, E-mail: pengdeliang@caas.cn.

at 24 h and 3 d rather than 8 d, and eight ROS-producing genes including three class III peroxidase and five lipoxygenase genes, were significantly up-regulated. Besides high throughput identification of candidate resistance genes, functional characterization indicated that phospholipase linking with ROS production played vital roles in early defense responses to *H. avenae* by involving diverse defense-related pathways as a hub switch. We are the first to investigate the incompatible defense responses of wheat against *H. avenae*, and our results not only provide applicable candidate resistance genes for breeding novel wheat cultivars, but also strengthen better understanding of defense mechanisms between cereal crops and cereal cyst nematodes.

Key words: cereal cyst nematode; anti-development; transcripts filteration; genetic background unification; expression profile; plant-nematode interactions

Recent advances and current status of heat treatment in fruit biocontrol system (host fruit-fungal pathogen-biocontrol agent)

Huizhen Chen, Jia Liu*

(School of Biotechnology and Food Engineering, Hefei University of Technology, 193 Tunxi Road, Hefei 230009, China)

Abstract: Eco-friendly approach like heat treatment and biological control utilizing antagonistic yeasts to manage postharvest diseases of fruits is a research topic that has been drawing considerable attention. The current review focuses on the effects of heat treatment on not only fruit commodities, but also fungal pathogen and biocontrol yeasts at molecular and physiological levels. The effects of postharvest heat treatment on aspects of defense response in fruit, by biochemical, transcriptional, proteomics and metabolic analysis, are discussed. Pathogenesis-related proteins, antioxidant enzymes, heat shock proteins, novel genes and antifungal compounds involved in the metabolism in response to heat are dealt with. On the other hand, heat treatment exhibits the direct inhibitory effect on fungal pathogens, and the modes of action like transcriptional regulation, oxidative injury and ATP consumption on pathogens have been proposed. More recently, mild heat stress adaptation in antagonistic yeasts to induce cross-protection against various stresses and enhance their biocontrol efficacy, are highlighted. An integrated management will more likely provide a viable alternative to synergetic fungicides for postharvest diseases, compared to a single approach. Combing a heat treatment of fruit hosts with biocontrol yeasts has shown enhanced control efficacy against postharvest diseases in various fruits. This review provides an overview of recent advances of heat treatment in postharvest system (host fruit-fungal pathogen-biocontrol agent), with specific emphasis on each component. It systematically contributes to a comprehensive understanding of the action mode of heat treatment, which is beneficial for guidance of further application.

作者简介:陈会珍,在读硕士生,主要从事水果病害的生防研究,E-mail:huizhen23@163.com。

* **通讯作者**:刘嘉,教授,主要从事植物病害的生物防治研究,E-mail:jialiu1983@163.com。

Candidate effector proteins of the necrotrophic apple canker pathogen *Valsa mali* can suppress BAX-induced PCD

Zhengpeng Li[1], Zhiyuan Yin[1], Yanyun Fan[1], Ming Xu[1], Zhensheng Kang[1,2], Lili Huang[1]*

([1]State Key Laboratory of Crop Stress Biology for Arid Areas, Northwest A&F University, Yangling 712100, China; [2]China-Australia Joint Research Centre for Abiotic and Biotic Stress Management, Northwest A&F University, angling 712100, China)

Abstract: Canker caused by the Ascomycete *Valsa mali* is the most destructive disease of apple in Eastern Asia, resulting in yield losses of up to 100%. This necrotrophic fungus induces severe necrosis on apple, eventually leading to the death of the whole tree. Identification of necrosis inducing factors may help to unravel the molecular bases for colonization of apple trees by *V. mali*. As a first step toward this goal, we identified and characterized the *V. mali* repertoire of candidate effector proteins (CEPs). In total, 193 secreted proteins with no known function were predicted from genomic data, of which 101 were *V. mali*-specific. Compared to non-CEPs predicted for the *V. mali* secretome. CEPs have shorter sequence length and a higher content of cysteine residues. Based on transient over-expression in *Nicotiana benthamiana* performed for 70 randomly selected CEPs, seven *V. mali* Effector Proteins (VmEPs) were shown to significantly suppress BAX-induced PCD. Furthermore, targeted deletion of *VmEP*1 resulted in a significant reduction of virulence. These results suggest that *V. mali* expresses secreted proteins that can suppress PCD usually associated with effector-triggered immunity (ETI). ETI in turn may play an important role in the *V. mali*-apple interaction. The ability of *V. mali* to suppress plant ETI sheds a new light onto the interaction of a necrotrophic fungus with its host plant.

Key words: apple Valsa canker; secreted protein; cell death; virulence factor; plant-fungus interaction

---

This study was financially supported by the National Natural Science Foundation of China (No. 31471732 and 31171796), the Program for Agriculture (nyhyzx201203034-03) and the 111 Project (B07049).

* Corresponding author: Lili Huang, E-mail: huanglili@nwsuaf.edu.cn.

EB1参与微管装配

Involvement of EB1 in microtubule assembly

朱原野，周裕军，周明国*
（南京农业大学植物保护学院，南京 210095）

Abstract: Plus end tracking proteins (TIPs) are a unique group of microtubule binding proteins that dynamically track microtubule (MT) plus ends. EB1 is a highly conserved TIP with a fundamental role in MT dynamics, but it remains poorly understood in fungal pathogen. According to the previous study, we propose that the alterations in α-and β-tubulin of fungal pathogen has changed the interaction between EB1 and microtubule which enhanced the resistance to drugs inhibit microtubule assembly. In order to figure out this problem, we expressed His-tagged EB1 and GST-tagged EB1 in *E. coli*. The two recombinant EB1 were purified by Ni and GSH sepharose affinity chromatography, respectively. The yields of two recombinant EB1 were 4 and 6 mg/L, respectively, with 90% purify. GST-Pull down demonstrated that no interactions between EB1 and tubulin monomers were detected. However, EB1 can enhance the polymerization of microtubule *in vitro*. Furthermore, the co-localization of EB1 and microtubule were observed by GFP-labeling technic and this result indicated that EB1 may bind to the polymerized microtubule. Taken all, these results are helpful for us to better understand the role of EB1 in microtubule assembly.

基金项目：国家自然科学基金项目（31201543）。

作者简介：朱原野，硕士研究生，E-mail：2014102133@njau.edu.cn。

* **通讯作者**：周明国，教授，主要从事杀菌剂毒理与抗药性研究，E-mail：mgzhou@njau.edu.cn。

一个含多 KH 结构域蛋白是稻瘟菌无性发育和侵染相关的形态建成所必需的

A multiple KH domain protein is required for asexual development and infection-related morphogenesis in the rice blast fungus

周威，潘嵩，阴长发，戚琳璐，杨俊*，彭友良

（中国农业大学植物病理学系，北京 100193）

Abstract：KH domain-containing proteins are widely conserved and play diverse roles in post-transcriptional gene control. However，few studies deal with their roles in pathogenesis of plant pathogenic fungi. Here，we identified a novel virulence gene *PCG6* that encodes a protein with five KH domains in the rice blast fungus *Magnaporthe oryzae*. Deletion of *PCG6* led to severely attenuated virulence toward host plants. Microscopic observation showed that the deletion mutant was reduced in appressorium formation，penetration peg formation，and development of invasive hyphae. In addition，the *PCG6* deletion mutant was also significantly reduced in vegetative hyphal growth and conidiation. Pcg6 encodes a cytoplasmic protein expressing in mycelia，conidia，appressoria，and infection hyphae. Moreover，Pcg6 was co-purified and physically interacted with several proteins involved in nonsense mediated mRNA decay (NMD). Interestingly，the expression level of *GLD1*，a gene for glutamate dehydrogenase，was markedly increased in the *PCG6* deletion mutant as compared with the wild type. Consistently，over-expression of *GLD1* in the wild-type strain exhibited similar phenotypes as the *PCG6* deletion mutant. Taken together，we proposed that the multiple KH domain protein Pcg6 regulates the glutaminolysis by NMD to control infection-related morphogenesis and developments in the rice blast fungus.

基金项目：国家自然科学基金项目（31371885）；国家"973"项目（2012CB114000）。

* **通讯作者**：杨俊，副教授，主要从事植物病原物致病机理、植物与病原物互作组学的研究。

水稻抗纹枯病分子育种新进展

Research highlight on rice molecular breeding toward sheath blight resistance

左示敏*，潘学彪，陈宗祥，张亚芳，陈夕军
（扬州大学，扬州 225009）

由强腐生性真菌“立枯丝核菌（*Rhizoctonia solani* Kühn）”引起的水稻纹枯病是水稻最重要的病害之一。历年的水稻病害测报数据显示（http://www.natesc.gov.cn/），纹枯病的发生面积和造成的产量损失始终位于水稻各病害之首。水稻对纹枯病的抗性受多基因控制，通过传统育种手段难以开展抗病育种。近年来，通过分子育种手段开展的抗纹枯病育种工作取得了一些新进展，有望加速抗病品种的培育进程。分子育种包括分子标记辅助选择育种和转基因育种两个方面。

分子标记辅助选择育种的前提是有可靠的抗纹枯病数量基因座（quantitative trait locus，QTL）。迄今已报道了多个抗纹枯病 QTL，但效应稳定且实现精确定位的并不多见。位于第 9 染色体长臂末端的 *qSB-9* 是目前普遍检测到的抗纹枯病 QTL，但在不同抗病亲本中有关该基因的定位区间却不尽相同，影响了其在抗病育种的应用。最近的研究显示三个不同抗病亲本在 *qSB-9* 上携带的是等位基因，同时实现了 $qSB\text{-}9^{TQ}$（供体亲本为特青/TQ）的精细定位（Zuo et al.，2014a，2014b）。通过标记辅助选择，发现导入 $qSB\text{-}9^{TQ}$ 可显著提高粳稻品种对纹枯病的抗性但不影响其农艺和产量性状（Chen et al.，2014）。除此之外，最近还新发现了 2 个效应稳定的抗纹枯病 QTL，并实现了它们的物理定位（Zhu et al.，2014）。这些进展为抗纹枯病分子标记辅助育种提供了可靠的基因资源，未来的研究将会侧重在这些 QTL 的聚合育种上（Chen et al.，2014）。

抗纹枯病转基因育种的最新进展也特别值得关注。纹枯病菌侵早期会分泌一系列水解酶降解植物细胞壁，如多聚半乳糖醛酸酶（Polygalacturonase，PG）和木聚糖酶等。水稻中存在抑制 PG 活性的蛋白（PG inhibiting proteins，PGIP），体外表达的 OsPGIP1 蛋白可抑制纹枯病菌 PGs 活性，超表达 *OsPGIP1* 可显著提高水稻对纹枯病的抗性（Wang et al.，2015）。此外，一般认为乙烯信号参与植物抵御腐生型病原菌的抗病反应。通过病原菌诱导型启动子驱动水稻乙烯合成途径中的关键基因 *OsACS2* 的表达，发现转基因水稻在病原菌侵染后内源乙烯合成明显增加，显著提高了水稻对纹枯病和稻瘟病的广谱性抗性，更值得注意的是转基因水稻的农艺和产量性状基本不受影响（Helliwell et al.，2013）。不同于以往利用外源基因提高抗病性的策略，这些研究均集中在改造和利用水稻自身基因上，这为进一步通过基因组编辑技术改造水稻自身基因培育非转基因抗纹枯病品种奠定了基础。

以上进展为水稻抗纹枯病分子育种储备了相关技术或基因资源，也为进一步研究水稻

* 通讯作者：左示敏，E-mail：smzuo@yzu.edu.cn。

抗纹枯病分子机理提供了重要基础，具有十分重要的意义。

参考文献

[1] Chen Z X, Zhang Y F, Feng F, *et al*. Improvement of *japonica* rice resistance to sheath blight by pyramiding *qSB*-9^{TQ} and *qSB*-7^{TQ}. Field Crops Research, 2014, 161:118-127.

[2] Helliwell E E, Wang Q, Yang Y N. Transgenic rice with inducible ethylene production exhibits broad-spectrum disease resistance to the fungal pathogens *Magnaporthe oryzae* and *Rhizoctonia solani*. Plant Biotechnology Journal, 2013, 11: 33-42.

[3] Wang R, Lu L X, Pan X B, *et al*. Functional analysis of *OsPGIP*1 in rice sheath blight resistance. Plant Mol Biol, 2015, 87:181-191.

[4] Zhu Y J, Zuo S M, Chen Z X, *et al*. Identification of two major QTLs, *qSB*1-1^{HJX74} and *qSB*11^{HJX74}, with resistance to rice sheath blight in field tests using chromosome segment substitution lines. Plant Dis, 2014, 98: 1112-1130.

[5] Zuo S M, Zhu Y J, Yin Y J, *et al*. Comparison and confirmation of quantitative trait loci conferring partial resistance to rice sheath blight on chromosome 9. Plant Dis, 2014a, 98: 957-964.

[6] Zuo S M, Zhang Y F, Yin Y J, *et al*. Fine-mapping of *qSB*-9^{TQ}, a gene conferring major quantitative resistance to rice sheath blight. Mol Breed, 2014b, 34:2191-2203.

Fusarium proliferatum 胞外分泌物及其致病性研究

焦铸锦,黄思良*

(南阳师范学院农业工程学院,河南省伏牛山昆虫生物学重点实验室,南阳 473061)

摘　要:玉米(*Zea mays* L.)是世界重要的粮食与饲料作物,在我国属第一大粮食作物。河南是我国主要粮食产区之一,玉米在河南省是最主要的秋季粮食作物。根据笔者前期调查结果,层出镰刀菌 *F. proliferatum*(Matsushima) Nirenberg 是引起河南玉米穗腐病的主要病原菌,该菌对玉米的为害不仅表现在产量上,更重要的是其产生的真菌毒素对玉米作为粮食及饲料的安全性构成严重威胁。目前尚无关于 *F. proliferatum* 胞外分泌物及其致病性的相关报道。研究 *F. proliferatum* 分泌的胞外酶及毒素等化合物与致病性的关系,对深入认识该病原菌致病机理和科学防控玉米穗腐病有重要意义。本研究以玉米穗腐病籽粒分离得到的层出镰刀菌菌株 Fp1 为供试菌株,分别以纤维素诱导培养基、果胶诱导培养基、淀粉诱导培养基和脂肪诱导培养培养病原菌,检测发现该菌可以产生淀粉酶、果胶酶和纤维素酶。其中果胶酶透明圈直径最大。菌株 Fp1 接入 Czapek's 液体培养基中(100 mL/250 mL),28℃,150 r/min 震荡培养 12 h。用 0.22 μm微孔滤膜过滤上清液,首先用 30 kD 超滤管 9 000 r/min 离心 20 min 滤除分子量较大蛋白,再用 3 kD 超滤管 9 000 r/min 离心 20 min 滤除分子量较小蛋白,获得分子量在 3 kD 以下的培养滤液。取 Czapek's 液体培养基,用 0.22 μm 微孔滤膜过滤后作为对照液。取 3 叶期玉米幼苗,用去离子水冲洗干净叶片。用毛细管管口轻压叶片。产生轻微的创伤后,在伤口处滴加一滴菌丝培养滤液(每组处理 10 株幼苗,做 5 组重复)。对照滴加一滴过滤后的 Czapek's 液体培养基。将处理后的玉米植株在光照培养箱中,28℃培养 12 h(湿度为 60%)。结果显示,胞外分泌物处理的叶片伤口水渍状斑面积大于对照,水渍状斑主要沿伤口边缘及叶脉方向扩展。由此推测胞外分泌物可能参与了 *F. proliferatum* 侵染的过程。

关键词:*Fusarium proliferatum*;玉米穗腐病;胞外酶;致病性

基金项目:河南省高校科技创新团队支持计划项目(2010JRTSTHN012)。

作者简介:焦铸锦(1978—　),男,博士,主要从事植物病害生物防治研究,E-mail:jiaozhujin@163.com。

* **通讯作者:**黄思良,E-mail:silianghuang@126.com。

*StpkaC*1调控玉米大斑病菌发育及致病性功能分析

张玉姝，申珅，张虹，郝志敏*，董金皋*

（河北农业大学真菌毒素与植物分子病理学实验室，保定 071000）

摘　要：玉米大斑病（northern corn leaf blight）是一种由大斑刚毛球腔菌（*Setosphaeria turcica*）引起的严重威胁玉米生产的真菌性病害。本课题组前期研究发现，cAMP 信号途径对玉米大斑病菌的营养菌丝生长及致病性具有重要的调控作用，PKA 作为该信号途径的下游调控因子，由两个催化亚基（PKA-C）和两个调节亚基（PKA-R）构成，针对该途径进一步从分子水平解析病原菌致病性的调控机制，将为寻找高效、持久的病害防控策略及发掘杀菌剂设计的新靶点奠定理论基础。本课题组前期利用 RNAi 技术获得沉默转化子，分析结果显示 PKA 调控病菌的附着胞发育及胞外水解酶活性影响其侵染能力，但其调控机制有待深入探索。因此，对玉米大斑病菌 *StpkaC*1 基因进行 PEG 介导的原生质体转化技术，经潮霉素抗性筛选、PCR、Southern bloting 验证得到 2 株敲除转化子。对 *StpkaC*1 基因敲除突变菌株的形态及致病力进行分析，结果显示，与野生型相比，突变菌株生长速率加快，菌落颜色变浅，气生菌丝较多，且出现产孢缺陷，附着胞形成及穿透时间延迟，黑色素产量明显低于野生型，且致病力下降。以上结果表明，*StpkaC*1 基因负调控菌株营养生长，对孢子产生、附着胞形成及穿透、黑色素合成均起正调控作用，并且调控菌株对寄主的致病力。

关键词：玉米大斑病菌；*StpkaC*1；附着胞发育；致病力

基金项目：国家自然科学基金项目（No. 31301616）；教育部高等学校博士学科点专项基金项目（No. 20131302120008）；河北省高等学校青年拔尖人才计划项目（No. BJ2014349Y）；河北省自然科学基金项目（No. C2014204111）。

* **通讯作者**：郝志敏，E-mail：hzm_0322@163. com；董金皋，E-mail：dongjingao@126. com。

北京地区10个快菜品种种传真菌的初步研究

卢志军[1],徐帅[2],蔡乐[3],胡学军[3],席昕[1],吴学宏[2]*

([1]北京市植物保护站,北京 100029;[2]中国农业大学植物病理学系,北京 100193;

[3]北京市昌平区植保植检站,北京 102200)

摘　要:本研究采用洗涤检验法和PDA培养基法对来自北京地区的10个快菜品种进行了种传真菌检测和初步鉴定。结果表明:10个快菜品种的干种子平均带菌率高达72.7%,其中8个品种的带菌率介于70.3%~100%,另外两个品种的带菌率低于10.0%;青霉属(*Penicillium* spp.)、曲霉属(*Aspergillus* spp.)、链格孢属(*Alternaria* spp.)、镰刀属(*Fusarium* spp.)真菌为干种子携带优势菌群,其中链格孢属真菌的平均携带率高达72.5%。无菌水洗涤后,10个快菜品种的种子平均带菌率为57.0%;青霉菌、曲霉菌、链格孢菌为水洗后种子携带的主要菌群,其中链格孢菌和青霉菌的平均带菌率分别为61.2%和29.7%。10个快菜品种中,从7个品种的种子洗涤液中检测到真菌,主要菌群为链格孢属、青霉属和曲霉属真菌,其中青霉菌和曲霉菌的分离比例较高。采用1% NaClO处理种子后,快菜种子真菌携带率介于4.0%~19.5%,平均携带率为9.1%,携带的真菌种类主要为链格孢菌、青霉菌和曲霉菌,其中链格孢菌的平均携带比例为92.0%。国内未见围绕快菜种传真菌的相关研究报道,本研究所取得的结果将为进一步明确快菜种传真菌的具体种类及其对快菜的致病性提供基础,为生产中预防通过种子进行传播病害以及有效地防治相关病害提供参考。

关键词:北京地区;快菜;种传真菌;检测;初步鉴定

基金项目:现代农业产业技术体系北京市叶类蔬菜创新团队。

作者简介:卢志军,硕士,高级农艺师,主要从事蔬菜病害综合防控研究,E-mail:13693683207@163.com。

徐帅,硕士研究生,主要从事蔬菜种传真菌检测研究。

卢志军和徐帅为共同第一作者。

* **通讯作者:**吴学宏,博士,副教授,主要从事作物真菌病害及其防治研究。

北京地区16个生菜品种种传真菌的初步研究

卢志军[1]，徐帅[2]，胡学军[3]，蔡乐[3]，席昕[1]，吴学宏[2]*
([1]北京市植物保护站，北京 100029；[2]中国农业大学植物病理学系，北京 100193；
[3]北京市昌平区植保植检站，北京 102200)

摘　要：本研究采用洗涤检验法和PDA培养基法对来自北京地区的16个生菜品种进行了种传真菌检测和初步鉴定。结果表明：16个生菜品种的干种子平均带菌率高达95.2%，其中14个品种的带菌率均为100%，另外两个品种的带菌率分别为25.0%和98.3%；青霉属（*Penicillium* spp.）、曲霉属（*Aspergillus* spp.）、链格孢属（*Alternaria* spp.）、镰刀属（*Fusarium* spp.）真菌为干种子携带的优势菌群，其中链格孢属真菌的平均携带率高达82.8%。无菌水洗涤后，16个品种的种子平均带菌率为89.1%；青霉菌、曲霉菌、链格孢菌、镰刀菌为水洗后种子携带的优势菌群，其中链格孢菌的平均带菌率为73.2%。16个生菜品种中，从11个品种的种子洗涤悬浮液中检测到真菌，主要菌群为链格孢属、青霉属和镰刀属真菌。采用1%NaClO处理种子后，16个生菜品种的种子携带真菌的平均比例为72.2%，其中11个品种的带菌率高达100%；携带的真菌种类主要为链格孢菌、青霉菌、曲霉菌和镰刀菌，其中链格孢菌的平均携带比例为75.6%。国内外围绕生菜种传真菌检测及相关的研究报道较少，本研究针对来自北京地区的16个生菜品种进行种传真菌的研究为国内首次，研究所取得的结果将为深入研究生菜种传真菌的种类及其为害、预防生菜种传真菌随种子进行远距离传播以及科学防治生菜生产中的病害具有一定的指导价值和借鉴意义。

关键词：北京地区；生菜；种传真菌；检测；初步鉴定

基金项目：现代农业产业技术体系北京市叶类蔬菜创新团队。

作者简介：卢志军，硕士，高级农艺师，主要从事蔬菜病害综合防控研究，E-mail：13693683207@163.com。徐帅，硕士研究生，主要从事蔬菜种传真菌检测研究。

卢志军和徐帅为共同第一作者。

* **通讯作者**：吴学宏，博士，副教授，主要从事作物真菌病害及其防治研究。

稻曲病菌的生命之谜

樊晶[1]，杨娟[1]，王宇秋[1]，李国邦[1]，赵志学[1]，黄富[2]，李燕[1]，王文明[1*]
([1]四川农业大学水稻研究所/作物重大病害重点实验室，成都 611130；
[2]四川农业大学农学院/生态农业研究所，成都 611130)

摘　要：稻曲病菌是一种侵染水稻颖花引起穗部发病的病原真菌。它特异性侵染水稻孕穗期的花丝，在颖壳内形成白色菌丝块并逐渐长大，撑开内外颖壳从合缝处外露，包裹整个小花，形成球状病原物，即稻曲球。稻曲球中含有危害人畜的真菌毒素，严重威胁着粮食生产安全。近年来，稻曲病在我国日趋严重，尤其是在水稻孕穗期遇上阴雨连绵天气的年份大规模流行。目前，关于稻曲病菌的生活方式和侵染机制有许多不明之处。例如，水稻孕穗期之前稻曲病菌如何生存？为什么水稻孕穗期的阴雨天气会加重病情？稻曲球的大小通常是水稻成熟籽粒的数倍，暗示其形成需要大量营养物质，那么稻曲菌如何从水稻颖花中获取营养形成稻曲球？我们研究发现，在潮湿有水条件下稻曲病菌能在多种田间杂草（如稗草、马唐、千金子等）叶片上附生生长，并在短时间内产生大量分生孢子。而且，从杂草叶片上分离的稻曲病菌可引起水稻发病。这一结果暗示稻曲病菌的附生特征可能为其在水稻孕穗期侵染颖花提供充足侵染源，也能解释为什么当水稻孕穗期赶上阴雨连绵的天气稻曲病发病更加严重。我们还发现，水稻颖花被稻曲病菌侵染后，其各个花器官被菌丝包裹，不能完成正常的开花过程，同时由于颖花授粉受精被阻断，子房不能发育灌浆。出乎意料的是，在感染了稻曲病菌的颖花中，具有胚乳特异性表达特征的水稻灌浆相关基因高度表达。由于这类基因与营养储存相关，推测稻曲病菌很可能通过激活水稻颖花的灌浆系统，并拦截灌浆过程中从源器官运输而来的大量营养物质，以供病原菌生长及稻曲球的形成。我们初步筛选到能激活水稻灌浆相关基因的稻曲病菌效应因子，目前正对候选效应因子进行转基因验证。

关键词：稻曲病；水稻颖花；灌浆；稻曲球；附生

基金项目：国家自然科学基金(31501598)。

作者简介：樊晶，副研究员，主要从事分子植物病理学研究，E-mail：fanjing7758@126.com。

* **通讯作者**：王文明，研究员，主要从事分子植物病理学研究，E-mail：j316wenmingwang@sicau.edu.cn。

稻瘟菌中一个致病新基因*PCG4*的功能研究

彭军波，杨俊，初宇，左玉山，彭友良*

（中国农业大学植物病理学系，北京 100193）

摘　要：由稻瘟菌引起的稻瘟病是威胁水稻生产的一大真菌病害，在水稻生产的苗期、叶期、抽穗期等多个时期均可发生，造成水稻产量的严重损失。目前关于稻瘟菌致病机理的研究报道与日俱增，但是关于稻瘟菌膜蛋白致病机理的研究却很少。在本研究中，通过插入突变、侧翼序列分离等方法，我们从稻瘟菌中鉴定了一个编码假定膜蛋白的致病新基因*PCG4*。与野生菌相比，*PCG4*敲除体的菌丝生长速率减少30%，近顶端细胞长度显著变短；分生孢子梗形态卷曲，顶端不产生分生孢子；丧失了对寄主植物水稻和大麦的致病性。亚细胞定位结果表明，Pcg4-GFP与内质网标记蛋白HDEL-RFP部分共定位，且在稻瘟菌发育的各个阶段均有表达。膜酵母双杂交结果表明，Pcg4与多个蛋白互作，包括3个膜蛋白和2个外泌蛋白。目前，作者正在对Pcg4互作蛋白的功能进行分析。

基金项目：国家“973”项目（2012CB114000）。

* **通讯作者**：彭友良，教授，主要从事病原真菌致病性的功能基因组、植物抗病遗传资源利用的研究。

毒素在高粱靶斑病菌致病过程中的作用

彭陈，葛婷婷，郭士伟

（江苏省农业科学院，农业生物技术研究所，江苏省农业生物学重点实验室，南京 210014）

摘　要：高粱靶斑病的病原菌为平脐蠕孢菌（*Bipolaris sorghicola*），它能够产生对动物、植物、真菌、细菌以及一些癌细胞都有不同程度抑制作用的毒素——蛇孢菌素。为了明确该毒素对病菌致病性的作用，本文利用蛇孢菌素生物合成途径的抑制剂洛伐他汀研究了抑制剂处理对高粱靶斑病菌致病性的影响。结果发现，高粱靶斑病菌的孢子经洛伐他汀处理后，对大麦叶片的致病性下降，而创伤接种和人为添加蛇孢菌素均可恢复其致病性；经洛伐他汀处理后，孢子对寄主细胞的侵染率明显下降，且侵染菌丝对邻近细胞的扩展不明显；而洛伐他汀对孢子及其萌发过程中的形态建成基本无影响。这说明蛇孢菌素在高粱靶斑病菌孢子的致病过程中起重要作用，并可能具有破坏寄主细胞结构的功能。

关键词：高粱靶斑病；平脐蠕孢菌；毒素；致病性

Effects of toxin on the sorghum target leaf spot pathogen pathogenicity

Chen Peng, Tingting Ge, Shiwei Guo*

(Provincial Key Laboratory of Agrobiology, Institute of Biotechnology, Jiangsu Academy of Agricultural Sciences, Nanjing 210014, China)

Abstract: The sorghum target leaf spot pathogen is *Bipolaris sorghicola*, which can produce ophiobolin. Ophiobolin is a host-specific toxin, and it can inhibit the growth of animals, plants, fungi, bacteria and some cancer cells. In order to clarify the impact of the toxins on pathogenicity, we use ophiobolin biosynthetic pathway inhibitors lovastatin to study the roles. The results showed that, the pathogenicity decreased in barley leaves, when the spores of *Bipolaris sorghicola* was treated with lovastatin, however, wounded leaf tissue inoculated with the treatment spores and artificially added ophiobolin could resume its pathogenicity. After treated with lovastatin, the efficiency of the spore infection decreased significantly, and the phenomenon of infection

基金项目：江苏省农业科技自主创新资金（CX(14)2001）；植物病虫害生物学国家重点实验开放课题（SKLOF201519）。

作者简介：彭陈，男，硕士，研究实习员，研究方向为真菌病害，E-mail：pengchen3316@163.com。

* **通信作者**：郭士伟，男，硕士，副研究员，主要从事真菌病害研究，E-mail：shiweiguo@jaas.ac.cn。

hyphae spread into neighboring cells was not obvious. Although lovastatin did not affect the morphogenesis of spore germination. This results indicate that, ophiobolin plays an important role in the pathogenic process of *Bipolaris sorghicola*, and the function may be to destroy the host cell structure.

Key words: sorghum target leaf spot; *Bipolaris sorghicola*; toxin; pathogenicity

对禾谷镰孢菌突变体的全基因组测序表明肌球蛋白-5的突变引起禾谷镰孢菌对氰烯菌酯的抗性

李斌，郑志天，刘秀梅，周明国

（南京农业大学植物保护学院杀菌剂生物学实验室，南京 210000）

摘　要：小麦赤霉病是由禾谷镰孢菌（*Fusarium graminearum*）引起的一种重要病害。氰烯菌酯是一种结构新颖，作用机制独特，且对镰孢菌有很好抑制效果的新型杀菌剂。该药剂与现有的多种不同作用机制的杀菌剂没有交互抗性，能够有效治理小麦赤霉病菌对多菌灵的抗药性问题。但在本实验室前期研究中发现，离体条件下禾谷镰孢菌极易对氰烯菌酯产生抗药性，抗药性风险高等至中等，不同于已有杀菌剂抗性机制。因此，探明氰烯菌酯抗性机制，对于制定和实施抗药性早期治理策略，确保小麦赤霉病的可持续控制具有理论指导意义。为了探索赤霉菌对氰烯菌酯的抗性机理，我们对抗性标准菌株 YP-1（由标准菌株 PH-1 通过氰烯菌酯诱导得到）进行测序与分析。基于高通量 Illumina 平台，读出 1 400 000 个序列，其中 92.8%能够与禾谷镰孢菌相关基因匹配。与 PH-1 相比，YP-1 包含了 1 989 个碱基突变导致 132 个基因发生突变。我们测定了其他禾谷镰孢菌（2021）以及相关抗性菌株的 22 个功能基因。这些抗性菌株均在编码肌球蛋白-5 的基因上发生点突变（编码 216、217、418、420、786 位的氨基酸）。通过同源双交换用抗性菌株的肌球蛋白-5 基因来替换敏感菌株 2021 的肌球蛋白-5 基因，得到的转化子对氰烯菌酯均表现出抗性。结果表明肌球蛋白-5 的突变与赤霉菌对氰烯菌酯的抗性有关。

番茄晚疫病菌和叶霉病菌对嘧菌酯和甲基硫菌灵的敏感性检测及抗药性风险分析

王秋实[1]，郑晖[1]，黄中乔[1]，朱春雨[2]，刘西莉[1*]

([1]中国农业大学，北京 100193；[2]农业部农药检定所，北京 100020)

摘　要：本研究对2013年采集自全国主要番茄产区的番茄病样进行分离、纯化，获得了番茄晚疫病菌472株和番茄叶霉病菌776株。通过室内毒力测定，建立了番茄晚疫病菌对嘧菌酯的敏感基线，并分别检测了田间番茄晚疫病对嘧菌酯、番茄叶霉病菌对甲基硫菌灵的抗性频率，进行了抗性风险分析。研究结果对明确供试的番茄两种主要病原菌的抗药性发展情况，科学评估病原菌的抗药性风险，以及制定有效的抗药性治理策略提供了重要参考。具体研究结果如下：

从来自山东和上海两省市未使用过嘧菌酯地区的发病番茄上分离获得62株晚疫病菌，测定其对嘧菌酯的敏感性。结果表明，嘧菌酯对供试菌株的平均 EC_{50} 为(0.088 ± 9.7)μg/mL。田间抗药性检测表明，从北京、上海、河北3个省市采集分离的田间菌株对嘧菌酯均表现为敏感，但从山东省番茄病样上分离获得的晚疫病菌中检测到敏感性下降的菌株，抗性频率为2.6%。同时，发现在田间以150 g/hm^2(有效成分))的相同剂量连续5次施用嘧菌酯后，敏感性下降群体表现为上升趋势，抗性分离频率达到12%。抗药性风险分析表明，晚疫病菌对嘧菌酯为中等抗性风险，建议制定科学的病害管理和抗性治理方案，避免田间晚疫病菌对醚菌酯抗性的发展和蔓延。

从来自河北和河南两省未使用过甲基硫菌灵地区的发病番茄上分离获得139株叶霉病菌，测定了其对甲基硫菌灵的敏感性。结果显示，河北、河南两省的番茄叶霉病菌已经普遍产生了抗药性，抗性频率高达95%。同时，检测发现来自河北、河南和辽宁三省的田间菌株中，92.14%为叶霉病菌抗性菌株，且均为高抗水平。抗药性风险分析表明，番茄叶霉病菌对甲基硫菌灵属于高等抗药性风险，不推荐继续使用甲基硫菌灵用于田间番茄叶霉病的防治。

关键词：番茄晚疫病；番茄叶霉病；杀菌剂；敏感性；抗药性风险

基金项目：农业部农产品质量安全监管(农药管理)农药品种风险监测项目；国家“863”计划项目(2012AA101502)资助。

作者简介：王秋实，男，农业推广硕士，E-mail：wqs0904@cau.edu.cn。

* **通讯作者**：刘西莉，女，博士，教授，E-mail：seedling@cau.edu.cn。

禾谷镰刀菌多菌灵抗药性的温度敏感性

周泽华

（南京农业大学植物保护学院农药科学系，南京 210095）

摘　要：由禾谷镰孢菌（*Fusarium graminearum*）引起的小麦赤霉病是世界上重要的小麦病害之一，给小麦生产造成了严重的损失。我国主要使用以多菌灵为代表的苯并咪唑类杀菌剂防治小麦赤霉病。实验室早期发现，温度会影响禾谷镰孢菌对多菌灵的抗性。本研究测定了 2013 年田间分离的 89 株对多菌灵具有不同抗性水平的禾谷镰孢菌菌株在不同温度下抗药性程度的差异，测定了 20 株高抗菌株的不同温度下的 EC_{50}。89 株菌株在低温下对多菌灵的抗性水平均会降低，20 株高抗菌株中，部分菌株在不同温度下 EC_{50} 有极显著差异，部分菌株差异较小。测定 β_2 微管蛋白基因定点突变菌株及 2021 在不同温度下的 EC_{50}，发现 EC_{50} 有不同程度的差异。Western 结果表明，低温下 β_2 微管蛋白表达量有不同程度的增加。

关键词：禾谷镰孢菌；多菌灵；低温；抗药性

禾谷镰刀菌腐生生长和侵染生长的细胞周期调控不同

Distinct cell cycle regulation during saprophytic and infection growth in *Fusarium graminearum*

江聪[1,2]，许金荣[1,2]，刘慧泉[1]*

([1]西农-普度大学联合研究中心/旱区作物逆境生物学国家重点实验室/西北农林科技大学植物保护学院，杨凌 712100；[2]美国普度大学植物及植物病理系，印第安纳州 IN47907)

摘　要：双相型(dimorphic)病原真菌如白色念球菌(*Candida albicans*)和玉米瘤黑粉菌(*Ustilago maydis*)等腐生阶段和致病阶段的细胞形态截然不同(酵母状态和菌丝状态)。已经证明细胞周期调控在从腐生向致病阶段的细胞形态转变中发挥着至关重要的作用。此外，半活体营养(hemi-biotrophic)植物病原真菌如稻瘟菌(*Magnaporthe oryzae*)和炭疽菌(*Colletotrichum* spp.)等虽然腐生和侵染阶段都以菌丝形态存在，但其在植物组织内产生的侵染菌丝呈球根状，明显不同于培养基上的营养菌丝形态，这种菌丝形态的改变与侵染阶段特异的细胞周期调控有关，但其具体的分子机制尚不清楚。细胞周期蛋白依赖性激酶 Cdc2 是细胞周期调控中最重要的因子，由于大多数病原真菌都只有一个 *CDC2* 基因，且是必需基因，导致该基因对腐生生长和致病生长的细胞周期调控作用难以研究。笔者课题组在研究中发现小麦赤霉病病原禾谷镰刀菌(*Fusarium graminearum*)拥有两个 *CDC2* 基因(*CDC2A* 和 *CDC2B*)，基因敲除研究证明二者在腐生生长和无性繁殖阶段功能冗余，可相互替代，但只有 *CDC2A* 在侵染生长阶段至关重要。进一步研究证明二者可自身互作并相互互作，可能以同源或异源二聚体形式发挥作用。序列比对和同源蛋白结构模拟发现 Cdc2A 和 Cdc2B 存在一些明显的序列和结构变异，可能与其功能差异有关。基因区段置换研究发现，Cdc2A 的 N-端和 C-端序列同时决定其在侵染生长中的调控作用。上述研究表明 Cdc2A 介导了侵染生长特异的细胞周期调控，今后需要进一步明确 Cdc2A 是如何介导这种特异的调控机制，从而揭示侵染生长细胞周期调控不同于营养生长的分子机理。

* **通讯作者**：刘慧泉，男，副研究员，主要从事病原真菌比较和功能基因组学研究，E-mail：liuhuiquan@nwsuaf.edu.cn。

环境因子对花生网斑病菌分生孢子萌发的影响

胡彦江[1]，金静[2]，鄢洪海[2]，张茹琴[2]*

([1]青岛农业大学生命科学学院，青岛 266109；

[2]青岛农业大学农学与植物保护学院，山东省植物病虫害综合防控重点实验室，青岛 266109)

摘　要：为了探明花生网斑病菌(*Phoma arachidicola* Marasas, Pauer & Boerema)对花生侵染致病的条件，本试验研究了环境因素包括温度、湿度、光照(黑暗、近紫光、全光照)对花生网斑病菌分生孢子萌发的影响。结果表明，在黑暗条件下，培养 5.5 h 时分生孢子开始萌发，24 h 时萌发率为 100%。分生孢子萌发率与温度之间的拟合模型符合二项式 $y=-0.570x^2+28.97x-277.1(R^2=0.808)$。据此模型，得出分生孢子萌发的最低温为 12.8℃，最适温为 25.4℃，最高温为 38.0℃。培养 24 h 时，水膜中分生孢子萌发率(35.4%)极显著大于空气相对湿度为 100%(8.8%)和 98%(0%)的萌发率。培养 24 h 时，黑暗(100%)和近紫光(97%)条件下的分生孢子萌发率极显著大于全光照下的萌发率(91%)。结论：分生孢子萌发的最有利条件是黑暗、温度为 25.4℃，且有水膜存在。

关键词：花生；*Phoma arachidicola*；分生孢子；萌发；环境因子

基金项目：青岛农业大学博士基金资助项目(1109301)；山东省“泰山学者”建设工程专项经费资助项目。

作者简介：胡彦江，实验师，主要从事植物学教学及科研工作，E-mail：yanjianghu@163.com。

* **通讯作者**：张茹琴，副教授，E-mail：zhruq-72@163.com。

我国黄瓜霜霉菌交配型的测定

张艳菊*,刘丽,刘欣欣,柴洪庆,高继杨
(东北农业大学农学院,哈尔滨 150030)

摘　要:由藻界卵菌纲古巴假霜霉菌(*Pseudoperonospora cubensis* (Berk. & M. A. Curtis) Rostovzev)侵染引起的黄瓜霜霉病,先后在全世界约70多个国家有发生,对黄瓜生产威胁极大。在适宜的环境条件下,传播流行速度快,涉及范围广,是黄瓜生产中的一种毁灭性病害。*P. cubensis* 为活体营养型的植物专性寄生菌,无性繁殖产生的孢子囊抗逆性差、寿命短、自然条件下难以越冬,而卵孢子能够抵抗不良环境条件,在病残体及病田土壤中越冬,成为第二年的初侵染源。卵孢子是在恶劣环境中存活最久的繁殖体,但黄瓜霜霉菌卵孢子在自然条件下很少见,截至目前对于黄瓜霜霉菌卵孢子性分化的研究报道较少。因此,本研究对我国黄瓜霜霉菌的交配型类型、分布范围等进行研究,为黄瓜霜霉病的防治提供理论依据。本研究采用直接配对法对离体黄瓜叶片进行点滴接种,测定2012—2014年在黑龙江、吉林、辽宁、北京、河北、山东、江苏、湖北和广东9个省16个黄瓜主产区采集的61株黄瓜霜霉菌的交配型。结果表明,黄瓜霜霉菌属于异宗配合,需要2个交配型同时存在才能产生卵孢子。61株菌株中,以A1交配型为主,共44株,占总菌株数的72.1%;17株A2交配型,占总菌株数的27.9%,这17株A2交配型为来自黑龙江省哈尔滨6株、吉林省长春1株、辽宁省沈阳2株、盘锦1株、北京2株、山东省寿光3株和临沂2株。将不同交配型菌株的孢子囊悬浮液以1∶1的比例混合接种于离体黄瓜叶片后,置于12.5℃、15℃、18℃、20℃、22℃和25℃的培养箱内培养,结果显示产生卵孢子数量最多的温度为18℃,在此温度下接种后6 d和7 d时卵孢子数量最多,但在25℃时不产生卵孢子。

关键词:*Pseudoperonospora cubensis*;异宗配合;交配型;卵孢子

基金项目:国家自然科学基金项目(31171792)。

* **通讯作者:**张艳菊,教授,博士生导师,研究方向为植物病原生物学,E-mail:zhangyanju1968@163.com。

利用 Real-time PCR 技术进行小麦条锈病潜育期叶片中菌量变化的监测

赵雅琼，谷医林，刘彬彬，温丽，马占鸿，王海光*

（中国农业大学植物病理学系，北京 100193）

摘　要：小麦条锈病是由条形柄锈菌小麦专化型（*Puccinia striiformis* f. sp. *tritici*）引起的一种主要发生在小麦叶部的病害，是我国乃至世界范围内小麦上的重要病害之一，流行时可造成严重的产量损失。本研究在室内将我国小麦条锈病菌的重要生理小种条中 33 号（CYR33）的夏孢子接种于培育好的高感小麦品种铭贤 169 幼苗上，接种后每隔 24 h 采集 60 片长势一致的小麦叶片，每两片作为一个样品，直到第 10 天潜育期结束，共获得 300 个样品，利用 Real-time PCR 技术测定潜育期小麦叶片 DNA 的绝对含量和小麦叶片中条锈病菌 DNA 的绝对含量，以实现对叶片中小麦条锈病菌病原的早期检测和菌量的动态监测。研究结果表明，小麦叶片 DNA 和条锈病菌 DNA 的扩增效率均较好，在小麦条锈病潜育期内，接种叶片内小麦条锈病菌 DNA 相对含量［条锈病菌 DNA 相对含量＝条锈病菌 DNA 量$_{ng}$×100％/（条锈病菌 DNA 量$_{ng}$＋小麦 DNA 量$_{ng}$）］不断增加，呈指数变化趋势，在潜育后期，病菌 DNA 含量增长较大，说明及早进行小麦叶片中条锈病菌的检测，对于该病害的监测和有效防治具有重要的实际意义。本研究利用 Real-time PCR 技术实现了条锈病菌的早期检测，并且实现了对小麦条锈病潜育期叶片中菌量的动态监测，有利于进行该病害的预测预报和及早防治。

关键词：小麦条锈病菌；潜育期；Real-time PCR 技术；病原检测；病害监测

基金项目：“973”计划项目（2013CB127700）；国家自然科学基金项目（31471726）资助。

作者简介：赵雅琼（1991—　），女，山西阳泉人，硕士研究生，主要从事植物病害流行和宏观植物病理学研究，E-mail：18700944409@163.com。

* **通讯作者：**王海光（1978—　），男，山东单县人，副教授，主要从事植物病害流行学和宏观植物病理学研究，E-mail：wanghaiguang@cau.edu.cn。

山东小麦赤霉病菌的种群组成及毒素化学型分析

高先悦,高淑敏,林森,于金凤*
(山东农业大学植物保护学院,泰安 271018)

摘　要:由镰孢菌引起的小麦赤霉病是一种世界性的病害。其不仅降低小麦产量、影响小麦的品质,并且病原菌产生的单端孢霉烯族(trichothecene)毒素对人畜的健康造成很大的危害。近年来,随着气候的变暖和免耕、轮作等耕作方式的推广,小麦赤霉病的发生及危害由长江中下游主病区逐年向北扩展。2012 年由于扬花期大面积降雨,导致山东小麦赤霉病大发生。因此亟待明确山东全省小麦赤霉病菌的种群组成和毒素类型,为大田防治提供理论依据。本研究于 2014—2015 年从山东 28 县市采集小麦赤霉病标本 300 余份,分离获得小麦赤霉病菌菌株 338 株。按照采集地点选取 120 株作为代表菌株,采用鉴定种和鉴定 B 型毒素化学型的特异性引物进行鉴定分析。致病种检测结果表明,山东小麦赤霉病菌均为禾谷镰孢菌,分为两种类型,分别为 *F. graminearum*108 株、*Fusarium asiaticum*12 株,其中 *F. graminearum* 占到 90.0%,为优势致病型;*Fusarium asiaticum* 是长江中下游麦区的优势致病型,过去在山东麦区未检测到。毒素化学型检测结果表明,Deoxynivalenol(DON)毒素化学型是山东小麦赤霉病菌主要毒素化学类型,供检测的 120 个菌株中 118 株属此种类型,将 DON 化学型进一步划分为 3-AcDON 和 15-AcDON,利用竞争性 PCR 检测其对应菌株分别为 17 株和 101 株;仅有 2 株病原菌产 Nivalenol(NIV)毒素化学型。小麦赤霉病菌的种群组成和毒素化学型存在一定关系,但并不完全对应。

关键词:山东小麦赤霉病;镰孢菌种群组成;毒素化学型

* **通讯作者**:于金凤,教授,E-mail:jfyu@sdau.edu.cn。

四川省小麦品种(系)对条锈病抗性评价及抗性基因的分子检测

龚凯悦,初炳瑶,王树和,马占鸿*

(中国农业大学植物病理学系,农业部植物病理学重点开放实验室,北京 100193)

摘　要:小麦条锈病(*Puccinia striiformis* f. sp. *tritici*)是四川省小麦生产上最重要的病害,因其暴发流行速度快,常造成小麦大面积减产乃至绝收。利用抗病品种是控制该病害最为经济有效的方法。为了解四川麦区近年小麦主栽品种对当前条锈菌流行小种的抗性水平;鉴定抗条锈病基因在该区小麦品种中分布状况,给该地区小麦安全生产与品种合理布局提供依据。以四川省小麦条锈菌当前流行小种 CYR32、CYR33 和 V26 菌系等比例混合接种到 100 个供试四川小麦品种(系),待其发病后调查普遍率和严重度及反应型,计算出各个品种的病情指数及不同品种的抗性水平;分别用 Yr10、Yr15、Yr17、Yr18 和 Yr26 基因有效的分子标记检测其在参试品种(系)中的分布状况。结果表明,供试 100 个品种中,对混合小种表现免疫或近免疫的品种有 19 个,占 19%,中度感病品种有 45 个,占 45%,高度感病品种有 36 个,占 36%;在 100 个品种(系)中,携带 Yr15 基因的频率为 23%,Yr17 基因的频率为 32%,Yr26 基因的频率高达 44%,但是没有鉴定出携带 Yr18 的小麦品种。

关键词:小麦;条锈病;抗病基因分子检测

基金项目:"973"项目(2013CB127700)。

作者简介:龚凯悦(1991—　),女,河北石家庄,硕士研究生,主要从事植物病害流行学研究,E-mail:gongkaiyue19911014@163.com。

* **通讯作者:**马占鸿,教授,主要从事植物病害流行和宏观植物病理学研究,E-mail:mazh@cau.edu.cn。

玉米大斑病菌漆酶基因 *StLAC4*、*StLAC6* 的功能研究

马双新，杨阳，刘宁，曹志艳*，董金皋*
（河北农业大学真菌毒素与植物分子病理学实验室，保定 071001）

The study of the functions of *StLAC4* and *StLAC6* in *Setosphaeria turcica*

Shuangxin Ma, Yang Yang, Ning Liu, Zhiyan Cao*, Jingao Dong*
(Mycotoxin and Molecular Plant Pathology lab, Agricultural University of Hebei, Baoding 071001)

摘　要：漆酶（Laccase，EC 1.10.3.2），也称为多铜氧化酶，其催化中心含有多个铜离子，能够氧化多酚、有毒芳香胺类等物质，广泛存在于真菌、细菌、昆虫和植物中。漆酶与色素合成、木质素降解、微生物菌体形态建成、子实体形成、病原菌致病性和脱毒作用等密切相关。玉米大斑病菌（*Setosphaeria turcica*）是导致玉米大斑病发生的丝状病原真菌。前期研究结果表明，玉米大斑病菌能够代谢产生漆酶，并且在致病过程中，漆酶可以促进病原菌在寄主组织中的扩展。本研究克隆获得了玉米大斑病菌 9 个漆酶基因，分属于不同的漆酶基因家族，利用基因敲除技术先后创制了其中漆酶亚家族 1、3 的 EOA82311（*StLAC1*）、EOA90070（*StLAC2*）两个漆酶基因的缺失突变体，通过对突变体表型及致病力分析发现，*StLAC1*、*StLAC2* 基因的缺失均降低了玉米大斑病菌漆酶活性、黑色素含量，菌丝形态及疏水性发生不同程度的变化，并且不能正常产生附着胞及分生孢子，使病菌丧失了传播及侵染能力。玉米大斑病菌漆酶基因 *StLAC4*、*StLAC6* 与 *StLAC1* 属于同一亚家族，在功能上可能存在互补作用，为了进一步验证漆酶基因在玉米大斑病菌发育、黑色素合成代谢及侵染过程中的作用，以及不同漆酶基因在功能上的差异，本研究以玉米大斑病菌基因组 DNA 为模板，通过 PCR 技术分别克隆获得了 *StLAC4*、*StLAC6* 基因的上、下游同源片段，并以质粒 pBARKS1 为骨架，草铵膦（Bar）和潮霉素（Hph）基因为抗性基因，成功构建了 *StLAC4*、*StLAC6* 基因的双交换敲除载体，拟创制 *StLAC4* 和 *StLAC6* 基因的缺失突变体，并通过有性杂交的方式获得 Δ*StLAC4*/Δ*StLAC6* 双突、Δ*StLAC1*/Δ*StLAC4* 双突、Δ*StLAC4*/Δ*StLAC6* 双突的菌株，最终明确 *StLAC1*、*StLAC4* 和 *StLAC6* 基因在玉米大斑病菌生长、代谢、致病等方面的作用及其基因间的功能互补关系。

基金项目：高等学校博士学科点专项科研基金（20111302120004）；河北省高等学校科学技术研究项目（Z2011109）。
* **通讯作者：**曹志艳，Tel：0312-7528142，E-mail：caoyan208@126.com；董金皋，Tel：0312-7528266，E-mail：dongjingao@126.com。

玉米大斑病菌中 GPCR 表达规律的研究

李贞杨，于波，申珅，刘宁，郝志敏*，董金皋*

（河北农业大学真菌毒素与植物分子病理学实验室，保定 071000）

摘　要：玉米大斑病是由玉米大斑病菌（*Setosphaeria turcica*）引起的玉米叶部重要病害，在流行年份常造成严重的经济损失。许多植物病原真菌的生长发育都受到细胞信号转导途径的调控，而 GPCR 即 G 蛋白偶联受体是细胞信号传导中的重要蛋白质，其拓扑构象为 7 次跨膜的受体。当膜外的配体作用于该受体时，该受体的膜内部分与 G 蛋白相互结合，激活 G 蛋白。G 蛋白可以激活第二信使如 cAMP 等，进而将胞外信号转化成胞内信号。因此，对 GPCR 的研究具有重要意义。本实验通过收集玉米大斑病菌野生型从孢子萌发、附着胞形成及成熟到侵染的整个过程中各个时期的材料，提取其 RNA，反转录得到 cDNA。进一步利用 Real-time PCR 技术，对 G 蛋白偶联受体基因在分生孢子发育形成侵染结构的过程中不同阶段进行相对表达量分析，结果表明，所有基因在附着胞成熟时期 12 hpi（hours post inoculation）均显著下调（$P<0.05$）；除 *StRtc*1 和 *Stfdd*123 以外，全部基因整体变化趋势均呈现上调/持平（3 hpi）——上调（6 hpi）——下调（12 hpi）——上调（24 hpi）的规律，尤其在侵入丝形成时期（24 hpi）显著上调（$P<0.05$）。在附着胞形成时期（6 hpi），仅 *StSte2p* 和 *StRtc*2 表达显著上调。*StRtc*1 和 *Stfdd*123 在各阶段中的表达量与对照差异不显著或显著降低。本研究为深入解析植物病原真菌 GPCR 超家族的功能提供了理论依据。

关键词：玉米大斑病菌；GPCR；Real-time PCR

基金项目：国家自然科学基金项目（No. 31301616）；教育部高等学校博士学科点专项基金项目（No. 20131302120008）；河北省高等学校青年拔尖人才计划项目（No. BJ2014349Y）；河北省自然科学基金项目（No. C2014204111）。

* **通讯作者**：郝志敏，E-mail：hzm_0322@163.com；董金皋，E-mail：dongjingao@126.com。

玉米抗病相关基因在玉米与丝黑穗病菌、黑粉病菌互作过程中的表达差异分析

The difference expression of the resistance-related genes between *Sporisorium reilianum* and *Ustilago maydis* interact with corn

邹晓威[1,2]，王娜[1,2]，夏蕾[1,2]，郑岩[1,2*]

([1]吉林省农业科学院，长春 130124；

[2]农业部东北作物有害生物综合治理重点实验室，公主岭 136100)

摘　要：提取玉米丝黑穗病与黑粉病发病叶片的 RNA，反转录获得 cDNA 后，通过玉米抗病相关基因登录号 AI881638、AW424529、CF028241、CO526016、CK371597、BM379188、AI649523、CF349132、BM074921、BM349111 的序列设计引物，对玉米抗病相关基因进行 RT-PCR 扩增，分析目标基因的表达差异。结果显示 BM379188、AW424529、AI881638、CF028241 在两种病害发生过程中均呈现上调表达，其他几个基因在侵染玉米叶片中均未检测到其表达，所选取的抗病相关基因在对照玉米叶片中均未检测到其表达。本研究为进一步明确两种病原菌与玉米的互作机制奠定基础。

关键词：玉米丝黑穗菌；玉米黑粉菌；抗病基因

基金项目：吉林省科技发展计划项目(20090710)。

作者简介：邹晓威(1983—　)，女，助理研究员，硕士，主要从事玉米-真菌互作相关研究，E-mail：zouxiaowei2008@126.com。

* **通讯作者**：郑岩，研究员，主要从事玉米-真菌互作相关研究，Tel：0431-87063867，E-mail：yanzhenga@yahoo.com。

玉米弯孢叶斑病菌(*Curvlaria lunata*)漆酶的致病性、基因克隆与黑色素合成相关性分析

吕宾，夏淑春，张茹琴，王麒然，鄢洪海*

（青岛农业大学农学与植物保护学院，青岛 266109）

摘　要：玉米弯孢菌叶斑病是玉米上的重要叶部病害，曾经造成过玉米严重减产，因此，研究其致病机制和建立行之有效的防控技术是生产上需要解决的关键问题。

本实验室之前的研究证实玉米弯孢叶斑病菌在活体外、离体叶片及愈伤组织中均可分泌一系列的细胞壁降解酶：多聚半乳糖醛酸酶(PG)、果胶甲基半乳糖醛酸酶(PMG)、漆酶、纤维素酶(Cx)和 β-葡萄糖苷酶。尤其是弯孢菌对玉米愈伤组织的侵染致病过程中，产生的漆酶活性是其他酶活性的 1.09～7.3 倍，且证实以木质素为诱导产生的细胞壁降解酶对寄主组织的浸解能力最强，引起了我们的特别关注。为此，我们进一步做了漆酶活性与弯孢叶斑病菌菌株致病力相关性分析，结果表明：弯孢叶斑病菌强致病力菌株 LNC0907 产生的漆酶活性显著高于弱致病力菌株 LNC1403，接种自交系黄早四植株后弯孢叶斑病的病情指数也存在显著差异，且漆酶活性与玉米弯孢叶斑病病情指数之间存在密切相关性(r=0.958)，以及抗性越弱的玉米品系诱导病菌漆酶活性提高也显著。

根据 GenBank 上已登录真菌漆酶基因氨基酸序列的保守区域设计兼并引物，成功克隆到玉米弯孢叶斑病菌强致病力菌株 LNC0907 的两个漆酶同源基因片段(593 bp)。推导其相应氨基酸序列进行比对分析，分别与棉花黄萎病菌(*Verticillium dahliae*)和尖孢镰刀菌(*F. oxysporum*)漆酶基因的序列同源性为 74%和 73%，为利用 RACE 技术克隆基因全长 cDNA 及进行基因的表达特性和功能鉴定等研究奠定了基础。

做玉米弯孢菌黑色素合成基因 *PKS*、*Brn*1 的 qRT-PCR 表达与病菌漆酶活性相关性，结果表明黑色素合成基因 *PKS*、*Brn*1 的瞬时表达与漆酶活性呈正相关，初步判断漆酶与玉米弯孢菌黑色素合成有关。

基金项目：山东省科技发展项目(2009GG10009022)；山东省自然科学基金项目(ZR2011CL005)；山东省“泰山学者”建设工程专项经费资助(BS2009NY040)。

作者简介：吕宾(1991—　)，男，硕士研究生，研究方向为植物病原真菌致病机制，E-mail：tpzhang@126.com。

* **通讯作者**：鄢洪海，教授，E-mail：hhyan@qau.edu.cn。

玉米大斑病菌菌丝体转录组分析

于波，李贞杨，申珅，刘宁，郝志敏*，董金皋*
（河北农业大学真菌毒素与植物分子病理学实验室，保定 071000）

摘　要：玉米大斑病(northern corn leaf blight)作为一种重要的真菌性病害，对玉米的生产造成了严重的经济损失。目前研究发现，由G蛋白介导的cAMP、MAPK以及Ca^{2+}信号转导途径对于植物病原真菌的生长、发育以及致病性等多方面发挥着重要的调控作用。PP2A是生物体内主要的Ser/Thr蛋白磷酸酶，位于信号转导通路的下游，并与多种蛋白激酶相互配合，通过磷酸化和去磷酸化的方式参与调节多条信号转导通路。为进一步明确玉米大斑病菌的作用机制，对大斑刚毛座腔菌野生型菌株进行了转录组测序分析。共得到了11 410 315对reads，总碱基数为2 281 618 354。GC含量为55.59%，质量值大于或等于30的碱基所占的百分比为87.96%。转录组数据与参考数据比对其比对效率为85.40%。使用SAMtools软件识别测序样品与参考基因组间的单碱基错配，查找基因区潜在的单核苷酸多态性(single nucleotide polymorphism, SNP)位点。SNP位点总数为2 813，基因区的SNP位点总数为1 516，基因间区SNP位点总数为1 297，转换类型的SNP位点数目在总SNP位点数目中所占的百分比为76.29%。使用Cufflinks软件，与基因原有的剪接模型进行比较，进行新可变剪接事件(alternative splicing events)预测。外显子跳跃事件数目为62，内含子保留事件数目为1 470，5′端可变剪接事件数目为55，3′端可变剪接事件数目为70。使用Cufflinks软件对Mapped Reads进行拼接，并与原有的基因组注释信息进行比较，共发觉589个新基因，使用BLAST软件将发掘的新基因与NR、Swiss-Prot、GO、COG、KEGG数据库进行序列比对，获得了新基因的注释信息。

基金项目：国家自然科学基金项目(No. 31301616)；教育部高等学校博士学科点专项基金项目(No. 20131302120008)；河北省高等学校青年拔尖人才计划项目(No. BJ2014349Y)；河北省自然科学基金项目(No. C2014204111)。

* **通讯作者**：郝志敏，E-mail：hzm_0322@163.com；董金皋，E-mail：dongjingao@126.com。

应用 Real-time PCR 鉴定四川省小麦品种(系)对小麦条锈菌的抗性差异

初炳瑶，龚凯悦，王树和，马占鸿*

(中国农业大学植物病理学系，农业部植物病理学重点开放实验室，北京 100193)

摘　要： 利用 Real-time PCR 检测 77 个四川省主栽小麦品种(系)潜育期叶片内条锈菌变化和相对含量的差异，按照分子病情指数(MDX)＝病原菌 DNA 量 (pg) /叶片总 DNA 量(ng)的计算方法计算分子病情指数，确定供试品种(系)对条锈菌的抗性水平。将四川省当前主要流行小种 CYR32、CYR33 和 V26 菌系等比例混合，均匀接种到不同品种(系)上，分别于潜育期第 1、5、10、14 天采样，提取小麦叶片和条锈菌 DNA，进行 Real-time PCR 检测。结果表明，在潜育期的 4 次采样中，第 1 天和第 5 天的分子病情指数与实测病情指数之间没有相关性，第 10 天和第 14 天的分子病情指数与实测病情指数之间有极显著相关性($P<0.01$)，第 14 天的相关性高于第 10 天，两者之间差异不显著。因此，Real-time PCR 方法可以作为传统品种抗病性鉴定的验证手段，用于小麦品种抗条锈水平检测。

关键词： 小麦；条锈病；Real-time PCR；潜育期

基金项目：“973”项目(2013CB127700)。

作者简介： 初炳瑶(1992—　)，女，山东青岛人，硕士研究生，主要从事植物病害流行学研究，E-mail：chubingyao@163.com。

* **通讯作者：** 马占鸿，教授，主要从事植物病害流行和宏观植物病理学研究，E-mail：mazh@cau.edu.cn。

新疆塔城地区植物锈菌分类的初步研究

王丽丽，玛丽卡·依玛木，日孜旺古丽·苏皮色来，李克梅*
（新疆农业大学农学院，乌鲁木齐 830052）

摘　要：锈菌是植物病原菌物中数量较大的一类病原菌，引起各种锈病，给人类农、林、牧业造成巨大经济损失。但近年来，国内外利用锈菌专性寄生性将它成功开发用于毒、杂草的生物防治，显示出锈菌与人类农业生产息息相关。

塔城地区位于新疆维吾尔自治区的西北角，地处东经 82°16′～87°21′、北纬 43°25′～47°15′，属温带大陆性干旱气候，辖五县二市，地形多变，具有草地、山地、河流、荒漠等多种生态环境，蕴藏着丰富的生物资源。

2014 年 6 月 25 日至 6 月 27 日、8 月 12 日至 8 月 20 日期间，本课题组在塔城地区进行锈菌调查采集，对 128 份植物锈菌标本进行形态学分类研究，共鉴定出锈菌 9 属 43 种，其中，国内锈菌新纪录 1 种：天蓝岩苣柄锈菌 *Puccinia mulgedii*（West.）Syd.；新疆锈菌新纪录 1 种：羊茅柄锈菌 *Puccinia festucae-ovinae* Tai.。涉及寄主植物 15 科 50 属 77 种，其中，国内锈菌寄主植物新纪录 19 种，新疆锈菌寄主植物新纪录 5 种。

初步区系分析表明，塔城地区植物锈菌以柄锈菌属 *Puccinia* 为主，有 24 种，占总数的 55.8%，为优势种群；其次为单胞锈菌属 *Uromyces*，8 种，占 18.6%；多胞锈菌属 *Phragmidium*，3 种，占 7.0%；胶锈菌属 *Gymnosporangium* 和栅锈菌属 *Melampsora* 分别为 2 种，各占 4.7%；长栅锈菌属 *Melampsoridium*、鞘锈菌属 *Coleosporium*、金锈菌属 *Chrysomyxa* 和糙孢锈菌属 *Trachyspora* 分别为 1 种，各占 2.3%。从寄主植物分析来看菊科植物 Asteraceae 有 14 属 25 种，占到寄主总数的 32.5%，为塔城地区植物锈菌主要寄主类群；其次是蔷薇科 Rosaceae 6 属 20 种，占 26.0%；禾本科 Gramineae 9 属 9 种，占 11.7%；藜科 Chenopodiaceae 3 属 5 种，占 6.5%；豆科 Leguminosae 3 属 3 种，占 3.9%；唇形科 Lamiaceae 2 属 2 种，占 2.6%；鸢尾科 Iridaceae、毛茛科 Ranunculaceae、杨柳科 Salicaceae、伞形花科 Umbelliferae、桦木科 Betulaceae、牻牛儿苗科 Geraniaceae、荨麻科 Urticaceae、凤仙花科 Balsaminaceae、百合科 Liliaceae、茜草科 Rubiaceae、大戟科 Euphorbiaceae、松科 Pinaceae 和蓼科 Polygonaceae 植物分别为 1 属 1 种，各占 1.3%。

本研究结果进一步丰富了新疆锈菌的物种多样性，为后续相关研究奠定理论基础。所有研究标本保藏于在新疆农业大学农学院植物病理系真菌标本室。

关键词：新疆；塔城地区；锈菌；分类鉴定

基金项目：国家公益性行业项目（农业）科研专项：草地病害防治技术研究与示范（201303057）。

作者简介：王丽丽，女，讲师，主要从事植物真菌病害研究，E-mail：1136862740@qq.com。

* **通讯作者**：李克梅，女，副教授，主要从事植物病理学研究，E-mail：835004213@qq.com。

小麦赤霉菌基因组高变区专化于植物侵染和病菌适应

王秦虎[1]，刘慧泉[1*]，王晨芳[1]，许金荣[1,2]

([1]旱区作物逆境生物学国家重点实验室/西北农林科技大学植物保护学院，杨凌 712100

[2]美国普度大学植物及植物病理系，印第安纳州 IN 47907)

摘　要：赤霉菌(*Fusarium graminearum*)引起的赤霉病，是小麦、大麦和玉米上最为严重的病害之一。到目前为止，高质量的基因组测序结果仅美国菌株 PH-1 一个。在本研究中，我们对中国小麦主产区的三个代表菌株进行了全基因组重测序。分析结果表明，所有测序菌株在序列和系统发育上彼此相距较远，中国菌株中每个菌株约有 100 个基因在功能上特异缺失。赤霉菌的变异位点分布不均，可分为两个亚基因组：包括一个高变的快速基因组和一个保守的慢速基因组。高变的快速基因组 GC 含量较低且基因长度较短，外显子数目变异较大。此外，该亚基因组中富集正选择基因、侵染上调基因和分泌蛋白基因。综上所述，我们的研究表明，赤霉菌存在两个不同变异速率的基因组且其快速基因组专化于植物侵染和病菌适应。根据上述研究，我们鉴定到了 106 个侵染相关基因。

关键词：小麦赤霉病；基因组变异；变异速率；正选择；RNA-seq；侵染相关基因

基金项目：西北农林科技大学基本科研业务费专项资金(2452015011)。

作者简介：王秦虎，男，博士后，主要从事病原真菌遗传和变异研究，E-mail：wangqinhu@nwafu.edu.cn。

* **通讯作者**：刘慧泉，E-mail：liuhuiquan@nwsuaf.edu.cn。

香蕉枯萎病原菌4号生理小种侵染特性的研究

肖荣凤，刘波，朱育菁，李燕丹

（福建省农业科学院农业生物资源研究所，福州 350003）

摘　要：香蕉枯萎病是由尖孢镰刀菌古巴专化型（*Fusarium oxysporum* f. sp. *cubense*）引起的维管束坏死的重要土传病害，严重威胁着世界香蕉种植产业。了解香蕉枯萎病原菌的侵染特性是该病害防治的重点之一。为了跟踪观察该病原菌在香蕉根、球茎以及假茎组织中的侵染与定殖特性，对尖孢镰刀菌古巴专化型4号生理小种菌株FJAT-3076进行了绿色荧光蛋白转化，获得遗传性状稳定的转化子FJAT-3076-T2。转化子FJAT-3076-T2与野生型菌株FJAT-3076的致病性无明显差异，接种24 d时，香蕉植株的发病率均为100%。利用激光共聚焦显微镜跟踪观察了菌株在香蕉植株体内的侵染情况。结果表明，接种后3～10 d，植株的假茎及叶片未表现任何症状，但少许须根出现褐变，此时可观察到病原菌以菌丝体或分生孢子的形式从幼根表细胞侵入，而且在褐变根组织内可观察到菌丝体的存在，但其他部位未检测到发荧光的菌丝；到第17天时，接种转化子及野生型菌株的植株下层叶片大部分黄化，球茎部有1/2的褐变，此时在根、球茎部及假茎部均可观察到发荧光的菌丝体；到第24天时，植株死亡，球茎部有3/4以上的组织褐变，此时在根、球茎部及假茎部均可观察到荧光的菌丝体，但叶部未观察到发荧光菌丝体。部分菌丝体扩展到假茎组织并在细胞间隙伸展，还有部分菌丝沿着木质部导管向上伸展。整个侵染过程，在根及假茎维管束组织中都可观察到分生孢子的存在。同时发现，在接种24 d时，香蕉的根、球茎及假茎组织的枯萎病菌的分布密度差异明显：在假茎部菌丝的密度最高，其次是根部，而球茎部的菌丝相当稀少。因此，若能有效阻止病原菌从球茎部侵染至假茎部，可推迟或降低该病害的发生。

关键词：香蕉；尖孢镰刀菌古巴专化型；侵染特性；绿色荧光蛋白

小蓟内生真菌的分离与鉴定

金静*，宋丽敏，路炳声，梁文星

（青岛农业大学农学与植物保护学院，山东省植物病虫害综合防控重点实验室，青岛 266109）

摘　要：小蓟是一种广泛分布的常见药用植物，本研究采用常规组织分离法对小蓟茎、叶及根的内生真菌进行了分离与鉴定，比较了不同消毒液对内生真菌分离的影响，分析了小蓟中内生真菌的种类、数量及内生真菌随时间动态在不同组织的分布情况。小蓟根、茎、叶组织先用75%乙醇浸泡 1 min 后，分别放置在 0.1%升汞溶液和 2%次氯酸钠溶液中，在 0.1%升汞溶液中处理时间超过 90 s 和在 2%次氯酸钠溶液中处理时间超过 6 min 时，均未长出内生真菌；而处理时间分别为 0～60 s 和 1～5 min 时，小蓟的根、茎、叶组织周围有内生真菌的生长，且种类和数量有所不同，说明消毒液的种类和消毒时间对内生真菌的分离有影响。在 6～9 月间，从小蓟根、茎、叶组织中分离的内生真菌经鉴定为 9 个属的真菌，包括青霉属(*Penicillium*)、曲霉属(*Aspergillus*)、镰刀菌属(*Fusarium*)、芽枝孢属(*Cladosporium*)、拟青霉属(*Paecilomyces*)、链格孢属(*Alternaria*)、漆斑霉属(*Myrothecium*)、木霉属(*Trichoderma*)和根霉属(*Rhizopus*)。其中，青霉菌(*Penicillium* spp.)和曲霉菌(*Aspergillus* spp.)的分离频率分别为 27.60%和 25.35%，是小蓟内生真菌中的优势菌，且在根、茎、叶中均有分布。小蓟在 6 月得到的内生真菌的种类及数量较少，7～9 月份得到的内生真菌种类及数量较多。小蓟叶片中内生真菌的种类和数量多于根和茎中的内生真菌的种类和数量。

关键词：小蓟；内生真菌；表面消毒；分离频率

基金项目：山东省“泰山学者”建设工程资助；山东省 2011 年高等学校科技计划(J11LC07)。

* **作者简介**：金静，女，博士，主要从事植物真菌病害及真菌多样性研究，E-mail：caroljin8100@163.com。

绿豆叶斑病菌分生孢子的形成及萌发条件研究

殷丽华，张海涛，柯希望，张盼盼，左豫虎*，郑殿峰
（黑龙江八一农垦大学，国家杂粮工程技术研究中心，大庆 163319）

摘　要：由变灰尾孢菌（*Cercospora canescens*）引起的绿豆叶斑病，又称红斑病，是我国及亚洲绿豆生产上危害最为严重的病害之一。国内外对该病的研究报道主要涉及田间防治药剂筛选，关于绿豆叶斑病病原菌的生长特性虽有研究，但对于该病菌分生孢子的产生条件以及分生孢子的萌发条件等的研究至今仍未见报道。

为明确绿豆叶斑病菌的产孢条件及萌发条件，本研究采用玉米粉培养基、马铃薯蔗糖琼脂培养基、马铃薯葡萄糖酵母浸膏琼脂培养基、高粱粒培养基和小麦粒培养基等 8 种培养基，筛选绿豆叶斑病菌的最适产孢培养基。试验结果表明，变灰尾孢菌的产孢量和产孢时间在不同培养基上表现差异明显，其中产孢最快、产孢量最大的为高粱粒培养基，于培养后 10 d 开始产孢，培养后 14 d 产孢量可达 9×10^5个/mL，相对于其他培养基而言，高粱粒培养基具有产孢快速、产孢量稳定等特点。除燕麦片培养基、察氏培养基和花生叶斑病尾孢菌培养基不产孢外，其余培养基均在培养后 12～14 d 开始产孢。温度对变灰尾孢菌产孢影响明显，25℃ 最有利于病菌产孢；光照条件对变灰尾孢菌的产孢影响不大。分生孢子萌发条件的研究结果表明，分生孢子在 25℃ 培养 3 h 即可萌发，培养 24 h 平均萌发率达到 88%。20℃ 培养 24 h 平均萌发率为 74%，30℃ 培养 24 h平均萌发率为 80%。25℃条件下，光照对孢子萌发影响显著，全黑暗有利于孢子萌发，培养后 20 h，萌发率可达 76.5%，光暗交替和全光照培养后 20 h，萌发率仅为 60% 和 37%。

本论文对绿豆叶斑病菌的最适产孢培养基、最适产孢条件以及最适的孢子萌发条件进行了系统研究，明确了变灰尾孢菌的最佳产孢条件及孢子萌发条件，为深入研究绿豆与叶斑病菌的互作提供前期基础，为病害防治工作提供重要的理论依据。

基金项目：国家科技支撑计划项目子课题（2014BAD07B05-H08）；“国家杂粮工程技术研究中心”组建项目（2011FU125X07）；中国博士后科学基金第 56 批面上资助项目（2014M561378）；校科研启动基金（2030010000）。

作者简介：殷丽华（1983—　），讲师，主要从事植物病理学研究，E-mail：yinlhua@163.com。

* **通讯作者：**左豫虎，教授，主要从事植物病理学研究，E-mail：zuoyhu@163.com。

炭疽菌侵染草莓植株病程发展不同阶段代谢组学分析

常旭念，胡志宏，代探，刘鹏飞*

（中国农业大学农学与生物技术学院植物病理系，北京 100193）

摘　要：草莓炭疽病(strawberry anthracnose)是世界草莓生产上的主要病害之一，主要危害匍匐茎、叶柄、叶片、根茎、花、果实等器官，可导致局部病斑和全株萎蔫，尤其在根上严重发生，导致草莓苗期的大量死株现象，造成严重经济损失。有关该病害的不同侵染阶段寄主组织内化合物的代谢水平变化鲜有报道。本文采用气相色谱-质谱方法，研究了炭疽病潜育期、显症初期和显症末期三个阶段叶片组织中小分子代谢物，了解炭疽菌侵染草莓植株引起的寄主代谢应答反应，并探索病原的致病机制。在温室接菌后3 d、7 d和14 d分别获得“无症”、“显症初期”和“显症末期”的草莓叶片及其健康组织对照，用于代谢组提取、衍生化及气相色谱-质谱检测。获得了草莓病样组织代谢组的指纹图谱，经数据库检索确认匹配度大于70%的代谢物共47个，其中包括了糖类、醇类和有机酸类等物质。主成分分析(PCA)分别将炭疽病潜育期、显症初期和末期病样区分为不同的组。与健康组织相比，方差分析显示病害侵染后病样组织中抗病相关物质红景天苷含量显著上升。此外，糖类物质大量积累，包括葡萄糖、阿拉伯糖、甘露糖、半乳糖、木糖、蔗糖、核糖、来苏糖和阿卓糖。而其他类物质含量部分上升或下降。潜育期与显症期样品比较，糖类物质中赤藓糖、果糖、葡萄糖、半乳糖在潜育期含量即变化，而阿拉伯糖、甘露糖、松二糖、核糖、来苏糖、阿卓糖含量在显症期发生显著变化，有机酸类物质多数在显症期含量发生显著变化。研究显示，炭疽病菌侵染诱导了寄主体内抗病性相关代谢物含量的积累，以及糖类、醇类、有机酸类代谢水平的上调，其代谢调整水平与病程发展密切相关，但是否为炭疽菌侵染特异性生物标志物有待于进一步研究阐明。研究可为应用代谢组学手段解析寄主与病原间互作提供参考。

关键词：草莓炭疽病；代谢组；气相色谱-质谱；潜育期；显症期

作者简介：常旭念，硕士研究生，植物病理学，中国农业大学农学与生物技术学院植病系，E-mail：changuxnian@cau.edu.cn。

* **通讯作者**：刘鹏飞，副教授，博士，植物病理学，中国农业大学农学与生物技术学院植病系，E-mail：pengfeiliu@cau.edu.cn。

温度、湿度对苹果疫腐病菌孢子萌发、侵染和潜育的影响

刘芳，李保华*

（青岛农业大学农学与植物保护学院，山东省植物病虫害综合防控重点实验室，青岛 266109）

摘　要：由恶疫霉菌（*Phytophthora cactorum*（Leb. et Cohn.）Schrot.）侵染所致的苹果疫腐病，除侵染苹果果实外，还为害苹果树的根颈部及枝干。该病属于偶发性病害，其发生与降雨、温度和土壤湿度关系密切，连续阴雨常导致严重发病，造成大量烂果，已成为值得高度重视的防治对象。为进一步明确苹果疫腐病的侵染条件、侵染时期，研制能够预测苹果疫腐病菌侵染时期和侵染数量的模型，本研究在人工控制条件下，测试了温度、湿度对恶疫霉菌游动孢子萌发、侵染和病害潜育动态的影响。初步结果表明，恶疫霉菌游动孢子在5～35℃下均能萌发，其中最适萌发温度在25℃左右。在自由水中，游动孢子在7个测试温度下的平均萌发率为40.24%，而在100%和99%的相对湿度下平均萌发率仅为8%。恶疫霉菌游动孢子在10℃、15℃、20℃、25℃下均能侵染富士苹果未成熟的果实，并导致果实发病。其中，25℃下侵染量最大，用浓度为1×10^4个/mL的游动孢子悬浮液接种果实，经过100%保湿培养7 d，每果平均侵染面积可达整个果实的70.95%。用疫腐病菌游动孢子接种的果实在20℃和25℃下培养4 d后可见明显的病斑，而在5℃和35℃下培养20 d仍无可见病斑。本研究结果可为苹果疫腐病的流行预测提供理论基础与数据。

基金项目：国家苹果产业技术体系（CARS-28）。

* **通讯作者：**李保华，教授，主要从事植物病害流行和果树病害研究，E-mail：baohuali@qau.edu.cn。

我国华北区甜菜苗期未知叶斑病害的诊断鉴定

李文兵，白进玲，王颖，张宗英，吴学宏，韩成贵*

（中国农业大学植物病理学系，北京 100193）

摘　要：2014 年 4 月底 5 月初，我国河北省和内蒙古两地甜菜种植区纸筒育苗的苗期甜菜叶片上出现一种新的叶部枯斑病害，症状为子叶和真叶上出现枯斑，中心黄白色，有紫色晕圈，一般在苗棚滴水处形成发病中心，仅河北省张家口地区张北糖厂纸筒育苗面积 14 万亩中约 2 万亩发生此病害，内蒙古集宁地区商都、察右前旗和赤峰地区林西等地均有发生。该未知的苗期病害会给甜菜的生产造成较大的经济损失。因此，有必要诊断鉴定出该未知病害及其病原。分别在张北和林西采集甜菜苗期叶部病害，利用马铃薯葡萄糖琼脂培养基（PDA）分离、培养、保存病原菌。结合分离菌落形态，提取发病叶片或分离的菌株菌丝 DNA，利用真菌检测的通用引物 ITS1 及 ITS4 检测结果得到一条单一条带，全长 541 bp，回收、连接 pMD19-T 载体和测序。无论是直接提取病叶 DNA 的 PCR 产物测序还是分离的病原菌提取菌落 DNA 测序得到的序列相同，NCBI 比对结果为极细链格胞菌（*Alternaria tenuissima*）。进一步利用引物 H3-1a 和 H3-1b 扩增分离物菌丝 DNA 结果得到一条大小 600 bp 左右单一条带，回收测序结果同真菌通用引物检测结果一致。采用无伤接种的甜菜叶片涂抹法进行致病性试验，7 d 后观察出现了相同的叶斑症状，并从接种后出现病害症状的叶片上分离检测确认了病菌。因此，从张北和林西采集的甜菜苗期叶部病害病原菌为链格孢菌，而不是一些技术人员认为的由甜菜尾孢菌所致病害。所获研究结果为有效控制此类病害、减轻甜菜损失提供理论指导。

致　谢：博天糖业股份有限公司张海和高鸿斌先生，荷兰安地公司北京代表处的秦树才先生协助采集甜菜样品。

基金项目：国家甜菜产业技术体系（CARS-2102）。

作者简介：李文兵，硕士毕业，主要从事甜菜病害诊断与控制研究，E-mail：liwenbinglwb@163.com。

* **通讯作者**：韩成贵，教授，主要从事植物病毒学与抗病毒基因工程，E-mail：hanchenggui@cau.edu.cn。

丝兰内生真菌的分离与鉴定

金静*，宋丽敏，张茹琴，梁文星

（青岛农业大学农学与植物保护学院，山东省植物病虫害综合防控重点实验室，青岛 266109）

摘　要：本实验采用组织分离法对青岛农业大学校园内的丝兰根、茎、叶组织进行内生真菌的分离，得到的结果为：在6～9月份，丝兰植物不同部位（根、茎、叶）的内生真菌共有12个属，包括青霉属（*Penicillium*）、曲霉属（*Aspergillus*）、镰刀菌属（*Fusarium*）、链格孢属（*Alternaria*）、丝核菌属（*Rhizoctonia*）、芽枝孢属（*Cladosporium*）、拟青霉属（*Paecilomyces*）、漆斑霉属（*Myrothecium*）、枝顶孢属（*Acremonium*）、柱隔孢属（*Ramularia*）、炭疽菌属（*Colletotrichum*）、聚端孢属（*Trichothecium*）及两类未产孢的真菌。其中，曲霉属的分离频率为32.9%，镰刀菌属的分离频率为23.1%，青霉属的分离频率为17.3%，属于丝兰内生真菌的优势菌属。不同部位的优势菌种类不同：根部的优势菌为镰刀菌属，其在根上的定殖率为38.7%；茎部和叶部的优势菌均为曲霉属，定殖率分别为28.0%和37.3%。另外，不同组织的内生真菌种类和数量都有差异，其中根上的内生真菌的种类和数量均最多，茎上的内生真菌数量最少，叶上的内生真菌种类最少。7月和8月出现的内生真菌的种类及数量较多，6月和9月较少。不同消毒液及消毒时间对内生真菌的分离有影响，将根、茎和叶组织预先都用75%乙醇消毒1 min，然后放置在不同消毒液中不同时间，根和茎组织在0.1%升汞液中消毒20 s，内生真菌的定殖率分别为66.7%和46.7%，而在2%次氯酸钠液中消毒2 min，内生真菌的定殖率可分别达到80.0%和53.3%；叶组织在0.1%升汞液中消毒30 s，内生真菌的定殖率为73.3%，而在2%次氯酸钠液中消毒3 min，内生真菌的定殖率可达到86.7%。

关键词：丝兰；内生真菌；分离鉴定；产孢诱导；定殖率

基金项目：山东省“泰山学者”建设工程资助；山东省2011年高等学校科技计划（J11LC07）。

* **作者简介**：金静，女，博士，主要从事植物真菌病害及真菌多样性研究，E-mail：caroljin8100@163.com。

水稻纹枯病菌 *RsPG* 基因的克隆表达及致病性分析

陈夕军，张家豪，张云，张璐，徐斌
（扬州大学园艺与植物保护学院，扬州 225009）

摘　要：多聚半乳糖醛酸酶是植物病原真菌重要的致病因子，也是病原菌接触寄主后产生的第一类水解酶。为明确多聚半乳糖醛酸酶在水稻纹枯病菌致病过程中的作用，以水稻纹枯病强致病菌株 YN-7 为研究对象，采用 PCR、RT-PCR 和 RACE 等方法克隆了 3 个多聚半乳糖醛酸酶基因。其中 *RsPG*2 ORF 为 1 633 bp，具 6 个内含子，编码区全长 1 311 bp(KP896519)，编码产物 436 个氨基酸，前 16 个氨基酸为信号肽序列；*RsPG*3 ORF 为 943 bp，包含 4 个内含子，编码区全长 720 bp(KP896520)，编码产物 239 个氨基酸，无信号肽；*RsPG*4 的 ORF 为 1 466 bp，包含 8 个内含子，编码区全长 1 038 bp (KP896521)，编码产物 345 个氨基酸，无信号肽。生物信息学分析表明，RsPG2、RsPG3 和 RsPG4 均具有真菌 PG 特有的保守结构域 NTD、DD、SHG 和 RIK。3 个 RsPG 的二级结构均以 α-螺旋、β-折叠和卷曲为基本结构单元，跨膜结构预测其均以从胞内向胞外分泌为主。将 3 个 *RsPG* 基因分别构建到真核表达载体 pPIC9K，转化受体毕赤酵母 GS115，经甲醇诱导表达，上清均具有多聚半乳糖醛酸酶活性；诱导 4～6 d 后，其最高酶活分别达 91.09 U/mL、91.09 U/mL 和 267.21 U/mL。采用针刺法将经柱层析纯化的真核表达产物接种至水稻叶鞘，48 h 后均能使水稻叶鞘出现明显的坏死斑。用表达产物处理四叶期水稻叶鞘，可使之产生浸解作用，24 h 后还原糖量（以 OD_{540} 值表示）分别为 1.061、1.053 和 1.664，细胞损伤率分别为 62.6%、55.5% 和 90.0%。Real-time PCR 分析表明，3 个基因在纹枯病菌侵染水稻过程中均能上调表达，且以接种后 48 h 表达量最高，但不同基因之间表达量差异极显著，说明这些基因均参与了病原菌对寄主的致病过程，但各基因作用大小不同。

关键词：致病性；多聚半乳糖醛酸酶基因；真核表达；水稻纹枯病菌

连续多代 UV-B 照射对小麦条锈病菌致病性的影响

刘彬彬，赵雅琼，温丽，马占鸿，王海光*

（中国农业大学植物病理学系，北京 100193）

摘　要：小麦条锈病（由 *Puccinia striiformis* f. sp. *tritici* 引起）是我国小麦生产中一种重要的气传流行性病害，该病害的发生流行受到多种因素的影响。紫外线可能对于小麦条锈病菌的变异、病害发生和传播等有所影响。本研究在实验室内，采用人工紫外灯照射的方法，在同一强度（150 $\mu W/cm^2$）的 UV-B 照射条件下，设定分别照射 0 min、30 min、60 min、90 min，对小麦条锈病菌夏孢子进行连续多代照射处理，研究了不同 UV-B 照射时间对小麦条锈病菌夏孢子致病能力的影响。选择高度感病小麦品种铭贤 169 进行条锈病菌繁育。接种后的铭贤 169 小麦苗在人工气候室内培育。结果表明，相同代数不同照射时间处理之间，随着 UV-B 照射时间的增加，条锈病菌致病性发生改变，夏孢子的萌发率逐渐降低，潜育期不断延长，侵染概率随之下降，病害严重度亦减少，产孢能力亦有所减弱。同一 UV-B 照射时间的不同代数之间，随着照射处理代数的增加，条锈病潜育期延长，严重度降低，病原菌侵染概率下降，产孢能力减弱。这些生物学效应随 UV-B 照射时间和代数的增加而有所加剧。结果表明，UV-B 照射时间增加，对小麦条锈病菌夏孢子的侵染能力、繁殖能力以及致病能力等有一定的抑制作用。这说明 UV-B 照射对小麦条锈病菌有多种生物学效应，并存在累积效应，为进一步研究小麦条锈病流行规律和条锈病菌致病性变异提供了重要参考。

关键词：小麦条锈病菌；UV-B 照射；适应性；流行学组分；气候变化

基金项目："973"计划项目（2013CB127700）；国家自然科学基金项目（31101393）。

作者简介：刘彬彬（1989— ），女，山东济宁人，硕士研究生，主要从事植物病害流行和宏观植物病理学研究，E-mail：liubbtoday@163.com。

* **通讯作者**：王海光（1978— ），男，山东单县人，副教授，主要从事植物病害流行学和宏观植物病理学研究，E-mail：wanghaiguang@cau.edu.cn。

西瓜蔓枯病菌对苯醚甲环唑的敏感性基线及抗性监测

刘顺涛，李雨，鲜菲，肖继芬，余洋，杨宇衡，毕朝位*

（西南大学植保学院，重庆 400715）

摘　要：西瓜蔓枯病是由瓜类黑腐球壳菌（*Didymella bryoniae*）引起的一种世界性的西瓜病害，在我国各西瓜产区均有发生，尤其在重庆贵州等西南地区是危害西瓜的最重要的病害，对西瓜生产造成严重损失，施用化学杀菌剂是防治该病害的主要手段。由于多菌灵等苯并咪唑类杀菌剂的常年使用而抗性普遍严重，近年来苯醚甲环唑等三唑类杀菌剂逐渐成为防治该病害的主要药剂。国内外对西瓜蔓枯病菌对苯醚甲环唑的抗药性研究报道极少，因此本文开展了西瓜蔓枯病菌对苯醚甲环唑的敏感性基线及抗性监测的研究工作。采用菌丝生长速率法测定了 101 株野生菌株对苯醚甲环唑的敏感性，其 EC_{50} 值为 0.013 48～0.170 61 μg/mL，最大 EC_{50} 值是最小值的 12.7 倍，不同敏感性菌株的频率呈连续单峰曲线近正态分布，平均 EC_{50} 值为（0.056 91±0.002 890）μg/mL，以此作为西瓜蔓枯病菌对苯醚甲环唑的敏感性基线。以苯醚甲环唑 15、50、100 μg/mL 作为区分剂量，室内平板测定了分离自重庆和贵州的西瓜蔓枯病菌 208 株对苯醚甲环唑的抗性，结果表明，两地的病原菌对苯醚甲环唑以敏感为主，占全部 97.12%，但在重庆武隆检测到 2 株低抗菌株，占当地菌株的 5.3%，在贵州息烽检测到 4 株中抗菌株，占当地菌株的 5.4%，未检测到高抗菌株。以上研究结果表明，在重庆贵州地区西瓜蔓枯病菌对苯醚甲环唑仍以敏感菌株为主，但在部分地区发现有抗性菌株的产生，因此在生产上应加强对该药剂的抗性监测，以指导合理用药。

关键词：瓜类黑腐球壳菌；苯醚甲环唑；敏感性基线；抗药性

基金项目：公益性行业（农业）科研专项（201303023）；重庆市自然科学基金（cstc2012jjA80035）；西南大学科技创新基金资助项目（Sz201202）。

作者简介：刘顺涛（1989—　），男，贵州息烽人，硕士研究生，主要从事杀菌剂方面的研究，E-mail：1169178884@qq.com。

* **通讯作者**：毕朝位，男，副教授，主要从事植物真菌病害及病原菌抗药性研究，E-mail：chwbi@swu.edu.cn。

效应因子 AVR-Pia、AVR1-CO39 及其结合区域 RATX1 的重组表达、纯化和晶体生长

张轶琨，郭力维，刘强，彭友良，刘俊峰*

（中国农业大学农学与生物技术学院植物病理学系，北京 100193）

摘 要：由稻瘟病菌引起的稻瘟病严重危害水稻的产量和品质安全。水稻与稻瘟病菌之间的互作符合“基因对基因”假说。目前，水稻的 23 个抗病基因和稻瘟病菌的 9 个无毒基因已经被克隆。已有的研究结果表明，水稻抗病蛋白 Pia 可以识别稻瘟菌两个非同源的效应因子 AVR-Pia、AVR-CO39 触发水稻的 ETI 反应从而引发水稻的过敏性坏死反应。Pia 由一对 NLR 蛋白 RGA5、RGA4 组成，RGA5-A 作为 RGA5 的功能转录本，其 C 端 RATX1 金属离子结合结构域直接与效应因子相互作用以识别这两种效应因子。本研究作者利用大肠杆菌表达系统将 Pia 的截短体 *RGA5A_S*(*RATX*1)分别与无毒基因 *AVR-Pia*、*AVR*1-*CO*39 进行重组表达，纯化以及晶体生长。为了获得高表达、高质量的目的蛋白，通过比较不同载体和表达菌株组合在不同诱导条件下目标蛋白的表达量，从中获得较优的表达诱导条件；在此基础上，利用不同层析技术、优化纯化条件如缓冲液，添加剂等因素，建立了以上两个串联重组表达蛋白的最优纯化流程。采用气相扩散法筛选和优化晶体生长条件，初步获得复合物 RGA5A-AVR-Pia 的微晶，有待进一步优化获得理想的可用晶体。本研究结果以期为分析具有 RATX1 结构域的 R 蛋白 Pia 如何识别效应因子激发防卫反应提供依据，进一步为持久抗病性的作物选育提供新的材料和理论基础。

关键词：稻瘟病菌；抗病蛋白；效应因子；蛋白纯化

基金项目：国家自然基金项目(31571990)。

作者简介：张轶琨，硕士研究生，E-mail：Yikun_zhang@126.com。

* **通讯作者**：刘俊峰，教授，E-mail：jliu@cau.edu.cn。

嗜热真菌 Beta-1,3-葡聚糖酶结晶、晶体结构解析与催化残基鉴定

Anastassios C. Papageorgiou[1]*，陈进银[2]，李多川[2]*

([1]芬兰图尔库大学，芬兰图尔库 20521；[2]山东农业大学，泰安 271018)

摘　要：真菌 Beta-1,3-葡聚糖酶(分为内切酶和外切酶)具有重要的生物学功能，涉及真菌的众多生长发育过程。尽管目前仅有真菌的一个内切 Beta-1,3-葡聚糖酶晶体结构被解析，但它的催化残基是推测性的，没有实验证明；另一方面，真菌外切 Beta-1,3-葡聚糖酶的晶体结构是不清楚的。本研究以嗜热毛壳菌(*Chaetomium thermophilum*)的 Beta-1,3-葡聚糖酶(CtGluN)为研究对象，将它的基因在毕赤酵母中获得高效表达，层析纯化蛋白并进行结晶，得到了它的晶体。X-衍射显示：酶蛋白以单分子不对称形式存在。定点突变实验表明：保守的 Q167、Q218、E608、E631 涉及酶的催化活性。产物鉴定显示：它是一种外切 Beta-1,3-葡聚糖酶。目前酶的晶体结构正在解析中。酶晶体结构解析和定点突变为理解真菌 Beta-1,3-葡聚糖酶的催化机制和热稳定性提供证据。

关键词：嗜热毛壳菌；*Chaetomium thermophilum*；Beta-1,3-葡聚糖酶；结晶；晶体结构

参考文献

Papageorgiou A C, Li D C. Expression, purification and crystallization of a family 55 beta-1,3-glucanase from *Chaetomium thermophilum*. Acta Cryst F, 2015, 71: 680-683.

基金项目：国家"863"计划(2012AA10180402)和国家支撑计划(2015BAD15B05)。

* 通讯作者

外源水杨酸诱导苹果对炭疽叶枯病的抗性研究

张颖，李保华，李桂舫，董向丽，王彩霞*

（青岛农业大学农学与植物保护学院，山东省植物病虫害综合防控重点实验室，青岛 266109）

摘　要：由炭疽菌（*Glomerella cingulata*）引起的苹果炭疽叶枯病（Glomerella leaf spot，GLS），是近年来在我国新发现的一种苹果病害，主要危害苹果叶片和果实，目前，对于该病的防控主要依靠化学药剂，而频繁使用化学杀菌剂不仅使果园生态环境恶化，给食品安全带来极大隐患，且该病发病迅速，潜育期短，病原菌一旦侵入寄主组织，几乎没有用药防治时间。利用植物诱导抗病性控制病害具有持效期长、抗病谱广及不污染生态环境等诸多优点，被认为是植物病害防治的一种新策略，水杨酸（SA）是植物体内重要的信号分子，对植物代谢过程起调控作用，现已发现水杨酸能诱导植物对多种病害产生抗性，在植物抗病过程中有非常重要的作用。因此，研究水杨酸诱导苹果炭疽叶枯病的抗性，将为利用植物诱导抗病性防治炭疽叶枯病提供理论依据。本研究以“嘎啦”（*Malus domestica* Borkh. cv. Gala）幼树一至两年生完全展叶枝条为材料，分别进行 SA/未接种处理、SA/接种处理、不加 SA/未接种处理、不加 SA/接种处理，在处理后 0 d、1 d、3 d、5 d 和 7 d 定期取样，测定叶片中防御酶活性及病程相关蛋白 *PR*1、*PR*5 基因以及几丁质酶、*β*-1，3-葡聚糖酶基因的相对表达量。结果显示，0.2 mmol/L SA 溶液的诱导抗病效果最明显；经 SA 预处理的叶片再接炭疽叶枯病菌，多种防御酶活性明显升高，且 SA 信号途径相关基因的表达量也明显升高。表明水杨酸可通过提高叶片中防御酶活性以及调节 SA 信号途径相关基因的表达量，而诱导‘嘎啦’苹果叶片对炭疽叶枯病的抗性。

关键词：炭疽叶枯病；水杨酸；防御酶；病程相关蛋白基因；植物诱导抗病性

基金项目：国家自然科学基金（31272001 和 31000891）；现代农业产业技术体系建设专项资金（CARS-28）；山东省科技攻关计划（2010GNC10918）。

作者简介：张颖，女，硕士研究生，研究方向果树病理学，E-mail：zying1017@126.com。

* **通讯作者**：王彩霞，女，教授，主要从事果树病害及分子植物病理学研究，E-mail：cxwang@qau.edu.cn。

油菜素内酯提高水稻抗病性的分子机制

王凤茹，高梦烛，高静，张昊，董金皋*

（河北农业大学生命科学学院植物分子病理学实验室，保定 071001）

摘　要：为探明油菜素内酯（BR）提高水稻（*Oryza sativa*）抗病性的分子机制，本研究比较分析了 0、1×10^{-6} mol/L BR 处理 6 h 的水稻幼苗叶片总蛋白的双向电泳图谱，发现了 5 个明显受 BR 调控的蛋白点，经 MALDI-TOF 质谱鉴定，受 BR 上调的蛋白是 OsCHRLK（chitinase-related receptor-like kinase）；通过 RT-PCR 技术对水稻叶片中编码几丁质酶（chitinase）的基因 *CHA*1 的表达量进行了分析，发现 *CHA*1 受 BR 的上调；对 BR 处理后水稻叶片细胞间隙液中几丁质酶的活性分析表明，1×10^{-6} mol/L BR 处理的水稻幼苗叶片细胞间隙内几丁质酶的活性是未经 BR 处理的 1.51 倍。因此，本研究结果表明，BR 处理可激活水稻体内几丁质酶信号系统，加速对真菌细胞壁的降解，降低真菌对植物的伤害。

关键词：油菜素内酯；水稻；抗病机制

基金项目：河北省自然科学基金项目（C2013204106）；河北省高等教育教学改革研究项目（2012GJJG058）。

* **通讯作者**：董金皋，E-mail：wfr15931945160@126.com，E-mail：dongjingao@126.com。

疏棉状嗜热丝孢菌转录组分析

耿志刚,陈进银,李多川[*]
(山东农业大学,泰安 271018)

摘　要:疏棉状嗜热丝孢菌(*Thermomyces lanuginosus*)是嗜热真菌中生长上限温度最高的真菌,广泛存在在自然界中,特别在生物质残体中。为了理解该真菌降解生物质的机制,本研究分别将在微晶纤维素上和葡萄糖培养 48 h 的疏棉状嗜热丝孢菌的菌丝体进行转录组测序与分析。转录组比较分析发现:有 4 485 个差异表达基因,其中上调基因 2 301 个,下调基因 2 184 个。上调基因中,log2. Fold_change 值大于 2 的 207 个,大于 3 的 95 个,大于 4 的 48 个,大于 5 的 25 个,大于 6 的 19 个,大于 7 的 14 个。显著上调表达的基因主要包括:糖苷水解酶、运转体、氧化酶、脱氢酶、转录因子和大量的假想蛋白等基因(hypothetical protein)。上述结果为进一步研究嗜热真菌降解纤维素的分子机制奠定了基础。

关键词:疏棉状嗜热丝孢菌;*Thermomyces lanuginosus*;纤维素酶解;转录组

参考文献

Berka R M, Grigoriev I V, Otillar R, *et al*. Comparative genomic analysis of the thermophilic biomass-degrading fungi *Myceliophthora thermophila* and *Thielavia terrestris*. Nat Biotechnol, 2011, 29: 922-927.

基金项目:国家"863"计划(2012AA10180402);国家支撑计划(2015BAD15B05)。

* 通讯作者

磷酸化对 MoSub1 与 DNA 结合活性的影响

易龙，赵彦翔，王珊珊，彭友良，刘俊峰*

（中国农业大学农学与生物技术学院植物病理学系，北京 100193）

摘　要：转录是基因表达的第一阶段，并且是基因调节的主要阶段。转录的顺利进行一般都需要转录辅助因子的协助。PC4 是介导上游激活子与通用转录机制相互作用的转录辅助因子。它能够和多种转录因子相互作用，参与许多基因的调控。PC4 是酵母 Sub1 同源的蛋白，它们能够结合 ssDNA，参与转录的起始、延伸、DNA 损伤修复以及活性氧的耐受等多种生理生化过程。该蛋白在稻瘟菌中的同源蛋白为 MoSub1，为了深入地了解 MoSub1 在稻瘟菌中调控转录的分子机制，作者着重研究了磷酸化对 MoSub1 的 ssDNA 结合活性的影响。研究表明，磷酸化对于酵母 Sub1 以及人源 PC4 的 DNA 结合活性有着不同的影响。实验室前期也鉴定了 MoSub1 的几个磷酸化位点，包括 S14、S17、S119、S151、S155 等。为了研究磷酸化作用如何影响稻瘟病菌中 MoSub1 与 DNA 的结合活性，作者将上述 5 个磷酸化丝氨酸位点突变为谷氨酸，构建了 MoSub1 5SE 突变体，用来模拟 MoSub1 的磷酸化状态。同时构建了 CKⅡ激酶表达载体，将 hCKⅡα催化亚基与 MoSub1 进行共表达，获得 CKⅡ激酶磷酸化的 MoSub1p，命名为 MoSub1 CKⅡ。首先作者使用各种层析技术（包括亲和层析、离子交换层析及凝胶排阻层析等）对各重组蛋白进行了纯化，并对纯化缓冲液组成（pH、离子强度及添加剂等）和层析策略等进行了优化以获得高纯度有活性的目的蛋白。然后，通过圆二色谱（CD）实验证明了 5SE 突变对 MoSub1 的二级结构没有明显的影响。利用等温滴定量热（ITC）实验检测了 MoSub1、MoSub1 5SE 和 MoSub1 CKⅡ与寡聚核苷酸 dT_{20} 的结合活性，结果表明它们的化学计量比均为 2，解离平衡常数 K_d 均在 0.4 μmol/L 左右。这表明磷酸化对于 MoSub1 与 ssDNA 的结合力没有明显影响，正在开展磷酸化对 MoSub1 ssDNA 结合的动力学影响的研究。

关键词：*Magnaporthe oryzae*；通用转录辅助因子；磷酸化

基金项目：“973”项目（2012CB114000）。

作者简介：易龙，硕士研究生，E-mail：longyi122790@163.com。

* **通讯作者：**刘俊峰，教授，E-mail：jliu@cau.edu.cn。

柑橘上两种检疫性疫霉的三重 PCR 分子检测

廖芳，刘跃庭，戴世泰，罗加凤

（天津出入境检验检疫局，天津 300461）

摘　要：冬生疫霉（*Phytophthora hibernalis*）和丁香疫霉（*Phytophthora syringae*）是柑橘属植株上寄生的两种重要检疫性真菌病害。针对两种致病菌，目前国内外已有形态学鉴定和危害的研究报道以及冬生疫霉病菌的 ITS 序列分析和 RFLP 初步研究，但未见有两种病原菌的同步分子检测的报道。本研究首先通过培养，观察比较了两种病原菌的菌落和形态特征。发现前者菌落呈玫瑰花瓣状，孢子囊呈长卵圆形或椭圆形，分生孢子为卵圆形；后者菌落呈菊花花瓣状，孢子囊呈宽卵圆形或倒梨形至椭圆形，分生孢子为单胞肾形。通过比对 GenBank 上公布的柑橘属水果上 15 株疫霉菌的 18S rDNA 基因序列、ITS 序列、heat shock protein 基因序列分别设计了疫霉的通用引物 18SF/18SR、冬生疫霉的特异性引物 GPHF/GPHR 和丁香疫霉特异性引物 GPSF/GPSR。在同一 PCR 管中 3 组引物（18SF/18SR、GPHF/GPHR 和 GPSF/GPSR）经优化体系后，18SF/18SR 扩增序列作为所有菌株菌丝基因组 DNA 提取模板的质量监控，针对冬生疫霉和丁香疫霉病菌单一模板的 18S 区域、ITS 区域和 heat shock protein 区域同时进行三重 PCR 扩增。利用此方法检测，冬生疫霉病菌出现 884 bp 的 18S 保守区扩增片段和 232 bp 的 ITS 基因的特异扩增带，丁香疫霉病菌出现 884 bp 的 18S 保守区扩增片段和 683 bp 的 heat shock protein 基因特异扩增带。所有菌株均未见非特异性片段的干扰。此三重 PCR 检测体系的建立为此两种检疫性真菌病害的同步分子检测提供了特异方法。

关键词：冬生疫霉；丁香疫霉；检疫性真菌；三重 PCR；特异分子检测

进境加拿大大麦中真菌病害的检疫鉴定

张莹，刘鹏，牛春敬，罗加凤
（天津出入境检验检疫局动植物与食品检测中心，天津 300461）

摘　要：随着人们生活水平的提高，对高品质啤酒的需求量也不断增长。进口大麦的质量直接影响酿造啤酒的等级，进境大麦中携带大量病残体，而不少病菌可通过种子和病菌残体远距离传播。因此，对进口大麦进行健康检测，提高酿造啤酒品质、防止危险性病菌传入我国，十分必要。本文从进境加拿大大麦的可疑种子及病残体中分离获得 21 个真菌菌株，对所分离的菌株进行培养性状及形态学观察，同时对核糖体 DNA 内转录间隔区（ITS 区）扩增和测序，鉴定出 6 个种，包括黑麦麦角菌（*Claviceps purpurea*）、梨孢镰刀菌（*Fusarium poae*）、燕麦镰刀菌（*Fusarium avenaceum*）、细极链格孢（*Alternaria tenuissima*）、大麦网纹病菌（*Pyrenophora teres*）、颖枯壳针孢（*Phaeosphaeria nodorum*）。本研究从进境大麦种子中检测鉴定出 5 属 6 种真菌，均为大麦致病性真菌。其中，麦角菌、镰刀菌、链格孢菌的次级代谢产物会产生真菌毒素，对动物和人体有不同程度的影响。我国每年进口大量的大麦，带菌情况较为复杂，做好进境大麦的检疫、监管工作，对保障啤酒工业安全生产、防止外来危险性病原菌在我国定殖、流行以致造成危害具有极其重要的意义。

关键词：大麦；真菌；检疫鉴定

小麦 metacaspase 基因 *TaMCA*1 的功能研究

郝影宾，王康，王晓杰，康振生*
（西北农林科技大学植物保护学院/旱区作物逆境生物学国家重点实验室，杨凌 712100）

摘　要：克隆得到小麦 *TaMCA*1 基因，*TaMCA*1 开放阅读框全长 879 bp，编码 292 个氨基酸，具有 L 型 metacaspase 基因典型的氮端结构域。实时荧光定量 PCR 结果表明，*TaMCA*1 在接种后 24 h、48 h 和 72 h 呈诱导上调表达趋势，其中在 48 h 其表达水平最高。其 GFP 融合蛋白表达分布于细胞质中，其 his 融合蛋白不具有 caspase 蛋白酶活性。*TaMCA*1 参与了双氧水诱导的酵母细胞死亡以及 Bax 诱导的烟草和小麦细胞程序性死亡的过程。*TaMCA*1 基因沉默后的组织学观察分析表明，与接种对照相比，表现在接菌后 48 h 菌丝长度明显减小，接菌后 120 h 菌落面积明显减小；24 h 活性氧浓度明显增强。结果表明，*TaMCA*1 可能是参与调控了细胞程序性死亡，在小麦与条锈菌致病过程中发挥了重要作用。

关键词：小麦条锈菌；半胱氨酸蛋白酶；实时荧光定量 PCR；瞬时过表达；基因沉默

* 通讯作者：康振生，E-mail：kangzs@nwsuaf.edu.cn。

液泡加工酶坏死相关基因的克隆及功能研究

王康，郝影宾，段小圆，康振生，王晓杰*
（西北农林科技大学植物保护学院/旱区作物逆境生物学国家重点实验室，杨凌 712100）

摘　要：植物为了抵抗病原菌入侵形成了一种复杂的免疫系统，包括快速、局部的细胞坏死-过敏性坏死反应（hypersensitive response，HR）。液泡加工酶（Vacuolar processing enzyme，VPE）具有类似半胱天冬酶（casapase）活性，半胱天冬酶是动物细胞中调控程序性死亡的关键控制开关。虽然在植物基因组中没有发现半胱天冬酶同源基因，但发现 VPE 可以通过催化成熟激活植物液泡中的水解酶而发挥类似于动物细胞中半胱天冬酶调控的 PCD 功能。为了研究液泡加工酶在小麦与条锈菌抗病互作 HR 中的作用，通过电子克隆获得新的小麦 VPE 基因，通过序列比对、进化树分析，命名为 TaVPE3。该基因编码的蛋白存在信号肽序列，未成熟的 TaVPE3 氮端区域相似性较低，而成熟蛋白的氮端区域则高度保守。进化树分析表明 VPE 家族成员与大麦亲缘关系较近，TaVPE3 为营养型 VPE。表达谱分析发现：TaVPE3 被机械伤害、干旱、水杨酸、乙烯和脱落酸显著诱导表达。TaVPE3 在小麦与条锈菌非亲和互作中有不同程度上调诱导表达，推测 TaVPE3 主要参与了小麦与条锈菌互作过程中的防卫反应。TaVPE3 在小麦、烟草叶片和酵母瞬时过表达实验表明，TaVPE3 本身不能诱发或抑制细胞坏死，但可增强不同胁迫诱发的细胞坏死，因此其对条锈菌无毒性小种侵染小麦过程中引起的细胞程序性死亡可能有不同程度的贡献。

关键词：小麦；条锈菌；液泡加工酶

* **通讯作者**：王晓杰，E-mail：wangxiaojie@nwsuaf. edu. cn。

坏死相关基因液泡加工酶的克隆及功能研究

王康，郝影宾，段小圆，康振生，王晓杰
（西北农林科技大学植物保护学院/旱区作物逆境生物学国家重点实验室，杨凌 712100）

摘　要：为了研究液泡加工酶(vacuolar processing enzyme，VPE)的功能，通过电子克隆获得新的小麦 VPE 基因，通过序列比对、进化树分析，命名为 TaVPE3。该基因编码的蛋白存在信号肽序列，未成熟的 TaVPE3 氮端区域相似性较低，而成熟蛋白的氮端区域则高度保守。进化树分析表明 VPE 家族成员与大麦亲缘关系较近，TaVPE3 为营养型 VPE。表达谱分析发现：TaVPE3 被机械伤害、干旱、水杨酸、乙烯和脱落酸显著诱导表达。TaVPE3 在小麦与条锈菌非亲和互作中有不同程度上调诱导表达，推测 TaVPE3 主要参与了小麦与条锈菌互作过程中的防卫反应。TaVPE3 在小麦、烟草叶片和酵母瞬时过表达实验表明，TaVPE3 本身不能诱发或抑制细胞坏死，但可增强不同胁迫诱发的细胞坏死，因此它对条锈菌无毒性小种侵染小麦过程中引起的细胞程序性死亡可能有不同程度的贡献。

关键词：小麦；条锈菌；液泡加工酶

鸢尾重花叶病毒全基因组序列分析

李永强*，尚巧霞，任争光，赵晓燕，魏艳敏，刘正坪

（北京农学院植物保护系，北京 102206）

摘　要：鸢尾是多年生的草本植物，既是观赏花卉又可以作为香料和药物的原材料。和其他多年生植物一样，鸢尾也容易被病毒侵染，比如鸢尾轻花叶病毒（*Iris mild mosaic virus*，IMMV），菜豆黄花叶病毒（*Bean yellow mosaic virus*，BYMV），水仙潜隐病毒（*Narcissus latent virus*，NLV），芜菁花叶病毒（*Turnip mosaic virus*，TuMV），虎眼万年青花叶病毒（*Ornithogalum mosaic virus*，OrMV）和鸢尾重花叶病毒（*Iris severe mosaic virus*，ISMV）。ISMV 是马铃薯 Y 病毒属成员，其部分基因组序列已经被克隆，但是还没有关于该病毒全基因组序列及基因组特征的报道。2013 年 8 月份，北京地区的鸢尾表现出花叶，坏死等病毒病害的症状。为明确引起该病害的病原，我们采用小 RNA 高通量测序技术对发病叶片样品进行鉴定。利用 Velvet 和 PFOR 组装出 784 条 contigs，其长度分布在 33～1 227 bp。通过 BLAST 比对分析对具有部分序列重叠的 contigs 进行重新组装，最终组装出 12 条 contigs。BLASTx 比对分析表明这些 contigs 与马铃薯 Y 病毒属成员编码蛋白的相似度较高，其中 CP 部分与 NCBI 上公布序列相似度达 97%，确定该病原为鸢尾重花叶病毒。设计特异性引物对缺失部分序列进行扩增，并最终组装出 ISMV 的全基因组序列。序列分析表明 ISMV 全长有 10 403 个碱基组成（去除 poly(A)序列），其 5′端和 3′端非编码区的长度分别为 123 bp 和 329 bp，编码一个 3 316 个氨基酸的多聚蛋白。大部分蛋白的剪切位点在马铃薯 Y 病毒属中都比较保守，但是在 NIb 和 CP 处比较独特，为 E/G。一些保守的氨基酸基序比如 P1 蛋白的^{352}H-8X-D-33X-S$^{3\ 951\ 506}$，CI 蛋白的 G-A-V-G-S-G-K-S-T$^{1\ 514}$和 $^{1\ 526}$V-L-L-I-E-P-T-R-P-L$^{1\ 535}$，NIa 蛋白的^{2350}H-34X-D-67X-G-X-C-G-14X-H$^{2\ 471}$，NIb 蛋白的$^{2\ 794}$C-D-A-D-G-S$^{2\ 799}$ and $^{2\ 859}$S-G-3X-T-3X-N-T-30X-G-D-D$^{2\ 902}$等在 ISMV 中均存在。CP 中与蚜虫传播相关的 D-A-G 模体在 ISMV 中突变为 T-A-G，HC-Pro 中与蚜虫传播相关的 K-I-T-C 突变为^{499}K-I-G-C^{502}。根据 NCBI 上公布的其他马铃薯 Y 病毒属成员的多聚蛋白序列，利用 MEGA 软件构建进化树，结果发现 ISMV 与洋葱黄矮病毒，水仙潜隐病毒，胡葱黄条病毒和 *Cyrtanthus elatus virus* A 构成一组，表明 ISMV 与这些病毒亲缘关系较近，是马铃薯 Y 病毒属的新成员。

* **通讯作者：**李永强，Tel：010-80799135，E-mail：lyq@bua.edu.cn。

荸荠感染两种 RNA 病毒的鉴定

张方鹏[1,2]，洪霓[1,2]，王国平[1,2]，王利平[1,2]*

([1]华中农业大学植物科学技术学院，湖北省作物病害监测与安全控制重点实验室，
武汉 430070；[2]华中农业大学，农业微生物学国家重点实验室，武汉 430070)

摘　要：荸荠属莎草科(Cyperaceae)荸荠属(*Eleocharis* R. Br.)一种营养价值较高的水生蔬菜。在生产中，荸荠一经感染病毒，通过球茎无性繁殖后代带毒，导致荸荠产量和品质的降低。目前，国内外对于荸荠病毒病的报道仅有 CMV 感染和一种 DNA 病毒(未发表)。本研究于 2014 年至今对湖北团风县及广西农科院资源圃的荸荠样品进行了疑似病毒病害调查，采集其叶状茎材料进行小 RNA 测序，通过数据分析发现样品中潜带两种新的正单链 RNA 病毒，暂且命名为 *Water chestnut virus* 1(WCV1)和 *Water chestnut virus* 2(WCV2)。

利用数据拼接得到的 contigs 设计引物对其序列进行扩增，获得的序列进行生物信息学分析，WCV1 基因组含有的开放阅读框分别编码甲基转移酶(Met)、解旋酶(Hel)、依赖 RNA 的 RNA 聚合酶(RdRp)、热激 70 蛋白(HSP70h)、外壳蛋白(CP)。基于 RdRp 和 HSP70h 氨基酸序列分析，WCV1 与长线形病毒科(Closteroviridae)其他成员的相似性分别为 30%～39%和 27%～37%；与 Closteroviridae 病毒其他几个属成员进化树分析，WCV1 独立为一个组群，推测 WCV1 为 Closteroviridae 新属的一个成员。另外，WCV2 基因组编码的开放阅读框主要有外壳蛋白(CP)、NTP 结合结构域(NTP-binding domin)、蛋白酶(Pro)、依赖 RNA 的 RNA 聚合酶(RdRp)；基于 RdRp 氨基酸序列分析，WCV2 与伴生病毒科(Secoviridae)矮化病毒属(*Waikavirus*)成员相似性最高，为 55%～57%，在系统进化中与该属其他成员聚集为独立分支，推测 WCV2 为 *Waikavirus* 的一个新成员。

此外，基于 WCV1 HSP70h 保守区域以及 WCV2 RdRp 区域设计引物对采集自湖北团风和广西农科院的样品进行检测。结果表明，湖北样品中携带有 WCV1，广西样品复合感染 WCV1 和 WCV2，且 WCV1 核苷酸序列分析表明来源于湖北和广西样品的相似性为 98%～99%。本研究首次明确了荸荠样品感染两种新的 RNA 病毒，相关内容还在进一步研究中。

关键词：荸荠；小 RNA 测序；Closteroviridae；Secoviridae；HSP70h；RdRp

基金项目：国家科技部支撑计划(2012BAD27B00)。

* **通讯作者：**王利平，副教授，研究方向为植物病毒学，Tel：027-87282498，Fax：+86-27-87384670，E-mail：wlp09@mail.hzau.edu.cn。

番茄褪绿病毒 RT-PCR 检测技术的优化及河南分离物的分子鉴定

冯佳[1]，孙晓辉[1]，高利利[1]，刘灵芝[1]，乔宁[2]，刘永光[2]，竺晓平[1*]

([1]山东省蔬菜病虫生物学重点实验室，山东农业大学植物保护学院，泰安 271018；
[2]潍坊科技学院蔬菜花卉研究所，寿光 262700)

摘　要：番茄是我国最主要的蔬菜之一，国内各地均有种植，随着栽培面积的不断扩大，病害发生的种类也不断增多。据报道，我国的春番茄每年因病毒病减产 30％以上，夏、秋番茄损失更为严重，有的年份或有的地块几乎绝收。番茄褪绿病毒(*Tomato chlorosis virus*，ToCV)隶属长线形病毒科(Closteroviridae)毛形病毒属(*Crinivirus*)，除大洋洲和南极洲之外，其他各洲均有报道。该病毒近几年在我国各地相继暴发流行，发病初期易与缺素症混淆，常常因误诊而延误防治，给番茄生产造成严重的经济损失。目前国际上对番茄褪绿病毒的检测主要采用 RT-PCR 技术，所用到的引物设计各不相同，针对的靶标序列也各有侧重，开发和优化高效、准确的分子检测方法，在植株发病的早期能进行及时预警，有利于控制该病害的发生和传播流行。鉴于此，参考已报道的番茄褪绿病毒 RT-PCR 检测技术体系，对番茄褪绿病毒的 RT-PCR 引物、反应条件等进行了比较、筛选和优化，得到了稳定性强、灵敏度和特异性均较高的引物组合和最优反应条件。应用优化的 RT-PCR 检测技术体系对河南设施蔬菜产区采集的疑似番茄病样进行检测，经 NCBI BLAST 比对发现，河南分离物的 *CP* 和 *HSP*70 序列与 GenBank 上传的 ToCV 典型分离物序列的同源性均达 97％以上，确定番茄褪绿病毒病在河南多个地区发生。同时通过对感病番茄样品中病毒粒子的提取和电镜观察，进一步证实河南设施番茄已感染 ToCV。本研究通过对番茄褪绿病毒 RT-PCR 检测技术的优化，建立了快速、稳定、灵敏度高的 ToCV 检测技术，为日后检测该病毒提供了参考依据，同时确定了河南设施番茄感染番茄褪绿病毒。

基金项目：国家公益性(农业)行业科研专项(201303028)；山东省科技发展计划项目(2014GNC111008)；山东省自然科学基金(ZR2012CM032)。

* **通讯作者：**竺晓平，教授，E-mail：zhuxp@sdau.edu.cn。

分析来源于病毒的小 RNA 深度测序数据挖掘新的核盘菌病毒

王前前，程家森，付艳苹，陈桃，姜道宏，谢甲涛*

（湖北省作物病害监测和安全控制重点实验室，华中农业大学；
农业微生物学国家重点实验室，华中农业大学，武汉 430070）

摘　要：真菌病毒（fungal virus 或 mycovirus）是一类感染真菌并在其中复制的病毒，在真菌中分布广泛，而且种类丰富。课题组前期分离得到一株形态异常核盘菌菌株 SZ-150，该菌株对油菜及其他寄主表现出弱毒性状，是一株典型弱毒菌株。提取菌株 dsRNA 并克隆其序列，先前确定菌株 SZ-150 中含有两种病毒（SsRV1 和 SsHV1）和一种卫星 RNA（SatH）。其中，病毒 SsRV1 与来自葱属葡萄孢的双节段病毒 BpRV1 属于同一个种的不同菌株，病毒 SsHV1 与感染板栗疫病菌的低毒病毒科（Hypoviridae）中的低毒病毒 3（*Cryphonectria hypovirus* 3）有很高同源性，而 SatH 是病毒 SsHV1 的卫星 RNA。为了验证菌株 SZ-150 是否仅仅含有以上病毒信息，对该菌株中来源于病毒的小 RNA 进行了深度测序，进一步对菌株中病毒产生的小 RNA 序列分析拼接。结果发现，除了获得来源于上述两种病毒及卫星 RNA 的小 RNA 信息外，还获得了来自另外一种未被报道的 RNA 病毒（暂且命名为 SsTV1）的小 RNA 信息。随后根据小 RNA 序列信息对该病毒进行序列拼接、克隆，获得了病毒 SsTV1 不完整序列。序列分析表明该病毒 3′端具有 Poly（A）结构，仅含有一个开放阅读框，编码 1 757 个氨基酸的多聚蛋白，含有甲基转移酶（methyltransferase）、解旋酶（helicase）以及依赖于 RNA 的 RNA 聚合酶（RdRp）三个保守结构域，与芜菁花叶病毒科（Tymoviridae）内的病毒具有低同源性，相似性仅为 26%左右。遗传聚类分析表明 SsTV1 隶属于芜菁花叶病毒目（Tymovirales），且与 Tymoviridae 的病毒在进化树中关系最近，但其独立成为一支。而且病毒 SsTV1 基因组结构与隶属于 Tymoviridae 的典型病毒有显著区别，因此，初步推定 SsTV1 是隶属于 Tymovirales 目的一个全新正单链 RNA 病毒，建议以该病毒为模式种建立一个新病毒科。利用原生质体再生技术，得到一株含有 SsTV1 病毒但致病力等表型正常的再生菌株，由此推断病毒 SsTV1 对寄主的致病力等生物学特性影响有限。对病毒 SsTV1 的分子特性及其在病毒中的进化关系正在进一步分析中。

基金项目：国家自然科学基金项目（31101398）；公益性行业（农业）科技专项（201103016）。

作者简介：王前前（1990—　），男，山东济宁人，博士研究生，主要从事分子植物病理学（真菌病毒方向）研究，E-mail：wqqwaityou@163.com。

* **通讯作者**：谢甲涛，副教授，Tel：027-87280487，E-mail：jiataoxie@mail.hzau.edu.cn。

河南省侵染番茄的两种双生病毒鉴定与针对性双重 PCR 检测技术体系的建立

刘灵芝[1]，孙晓辉[1]，冯佳[1]，高利利[1]，竺晓平[1*]，谢丙炎[2]

([1]山东省蔬菜病虫生物学重点实验室，山东农业大学植物保护学院，泰安 271018；

[2]中国农科院蔬菜花卉研究所，北京 10008)

摘　要：双生病毒是世界范围内广泛发生的一类具有孪生颗粒形态的植物单链 DNA 病毒，至今已在多个国家和地区的多种作物上造成毁灭性危害。2014 年 8 月份从河南省郑州市中牟县、新乡市、焦作市马村区、安阳市汤阴县保护地栽培番茄植株上采集到 37 份表现严重矮化、黄化和曲叶症状的疑似感染双生病毒的样本。利用双生病毒简并引物扩增得到约 535 bp 的特异片段，序列经过初步分析后再分别利用中国番木瓜曲叶病毒(*Papaya leaf curl China virus*，PaLCuCNV)全长引物和番茄黄化曲叶病毒(*Tomato yellow leaf curl virus*，TYLCV)特异引物进行检测，分别得到 2 749 bp 和 833 bp 的特异核苷酸序列。以 PaLCuCNV 全长引物扩增的片段，经测序后 BLAST 比对发现，河南省郑州市中牟县的五个分离物以及焦作市马村区、安阳市汤阴县的各一个分离物与 GenBank 已登录的中国番木瓜曲叶病毒焦作分离物(登录号：JX555979)最为接近，核苷酸一致率为 99.40%，确定上述地区设施番茄被 PaLCuCNV 侵染，发病范围有所扩大。37 份样品以 TYLCV 全长引物扩增，均能得到特异性的片段，经测序后与 GenBank 已登录的其他 TYLCV 分离物的序列核苷酸一致率为 99.04%。同时利用双生病毒卫星分子 DNAβ 特异引物对上述样品进行 PCR 扩增、克隆和测序，在焦作市马村区 TYLCV 阳性样品中扩增到 DNAβ 分子，全长为 1 331 bp，上述结果表明河南设施番茄中 TYLCV 普遍发生，且部分伴随卫星 DNAβ，局部地区有 PaLCuCNV 发生侵染为害且多与 TYLCV 混合侵染。同时，根据 PaLCuCNV，TYLCV 和卫星 DNAβ 已发表的序列设计 PCR 扩增引物，建立了特异性检测 PaLCuCNV，TYLCV 和 DNAβ 的单一、双重 PCR 体系并对引物的灵敏度进行了检测。目前，TYLCV 在山东、河南、河北、天津、广西、广东种植区已大面积暴发，严重影响农业生产。PaLCuCNV 近几年呈蔓延流行趋势，且在市场上并没有抗 PaLCuCNV 品种。对不同地区 TYLCV 和 PaLCuCNV 分离物的全基因组序列测定，分析 TYLCV 和 PaLCuCNV 分离物的分类归属和种群遗传变异情况，推测我国 TYLCV 和 PaLCuCNV 发生严重的原因，以及预测 TYLCV 和 PaLCuCNV 在今后的存在和发生情况，为各省防治 TYLCV 和 PaLCuCNV 提供理论基础。

基金项目：国家公益性(农业)行业科研专项(201003065)；山东省自然科学基金(ZR2012CM032)。

* **通讯作者**：竺晓平(1966—　)，男，教授，主要从事植物病理学研究，E-mail：zhuxp@sdau.edu.cn。

黄瓜花叶病毒诱导的基因沉默载体用于玉米基因功能研究

王蓉,杨新鑫,王廿,刘学东,范在丰,周涛*

(中国农业大学植物病理学系,农业部植物病理学重点实验室,北京 100193)

摘　要:病毒诱导的基因沉默(Virus induced gene silencing,VIGS)广泛用于植物基因功能研究,已有多种病毒/病毒卫星开发为 VIGS 载体,用于重要作物基因功能研究。目前有报道雀麦花叶病毒(*Brome mosaic virus*)可用于玉米基因沉默,但仅能用于 Va35 等少数不常用自交系,且该病毒在玉米上引起的症状严重。我们将自然条件下侵染玉米的黄瓜花叶病毒北京分离物(ZMBJ-CMV)基因组进行改造构建了 ZMBJ-CMV VIGS 载体,成功沉默了玉米 B73 中的 *ZmPDS*、*ZmIspH*、*ZmChlI* 基因,沉默效率达 62%以上。ZMBJ-CMV-$2b_{N81}$::$ZmIspH_{215}$ 接种 B73 后 18～60 d,玉米第 1～5 片真叶中的 *IspH* 均可被有效沉默(60%～85%)。插入 ZMBJ-CMV-$2b_{N81}$ 载体的 *ZmIspH* 的片段大小为 103 bp 和 215 bp 时能引起玉米叶片中 *IspH* 的沉默(75%～80%)。ZMBJ-CMV-$2b_{N81}$::$ZmIspH_{215}$ 接种 24 种中外常用玉米自交系后发现其中 18 种自交系中的 *IspH* 可被沉默(42%～92%)。本研究成功构建了可有效沉默玉米基因的 ZMBJ-CMV VIGS 载体,该载体具有病毒症状轻微、沉默持续时间长、可侵染玉米自交系范围广等显著优点。

关键词:VIGS;黄瓜花叶病毒;玉米;基因功能研究

基金项目:国家自然科学基金项目(31371912)。

作者简介:王蓉(1989—　),女,博士研究生,E-mail:wangrong1019@163.com。

* **通讯作者**:周涛(1978—　),副教授,主要从事植物病毒及其致病机制研究,E-mail:taozhoucau@cau.edu.cn。

利用小 RNA 深度测序对我国北方地区甜菜病毒病的调查

仇心钰，郝丹丹，韩成贵，王颖*

（中国农业大学植物病理学系，农业生物技术国家重点实验室，北京 100193）

摘　要：甜菜（*Beta vulgaris*）是目前世界上主要的糖料和能源作物之一，具有较强的耐盐碱能力，在我国北方的农业生产中具有重要地位。病毒病是当前制约甜菜产业发展的一个重要的因素。目前常规的病毒病的诊断技术包括电子显微镜观察、鉴别寄主接种、血清学和分子生物学检测等，但常规方法对那些未知的病毒或者含量较低的病毒很容易造成漏检。近几年发展的第二代测序技术（next generation sequencing，NGS）是一种能一次对几十万到几百万的 DNA 分子进行序列测定的高通量的测序技术，因此又被称为深度测序（deep sequencing）。本研究利用深度测序的方法对我国北方甜菜样品进行检测，明确甜菜上病毒发生的种类和优势群体，同时试图发现之前用 ELISA 和 RT-PCR 所未能鉴定出的新病毒。

我们于 2014 年甜菜收获季节期间采集了新疆、内蒙古、甘肃和黑龙江省的甜菜样品，采集到的甜菜其叶部症状主要有花叶和畸形两大类，而根部则表现为健康或者不同程度的丛根症状。对甜菜样品按叶部和根部取样，参照 NEB 小 RNA 文库制备（Next® Multiplex Small RNA Library Prep Set for Illumina）说明书分别构建了叶部和根部的两个 cDNA 文库池，由北京诺禾致源公司进行测序，利用拼接软件 Velvet 对小 RNA 进行拼接，将组装的 contig 在 NCBI 上进行比对分析。结果显示在叶部文库中检测到甜菜花叶病毒（*Beet mosaic virus*，BMV），在根部文库中检测到了甜菜坏死黄脉病毒（*Beet necrotic yellow vein virus*，BNYVV）、甜菜神秘病毒 1 号（*Beet cryptic virus* 1，BCV1）和甜菜神秘病毒 2 号（*Beet cryptic virus* 2，BCV2）。进一步将分析不同地区来源病毒的遗传变异，并利用 RT-PCR 技术获得甜菜神秘病毒 1 号和 2 号的基因组全序列，为甜菜病毒病的检测与防控提供参考数据。

基金项目：中国农业大学 2015 年大学生创新计划项目（201510019027）和国家自然科学基金（31401708）部分资助。

作者简介：仇心钰，植保专业 2013 级本科生，E-mail：qiuxinyu.cau@foxmail.com。

* **通讯作者：**王颖，讲师，主要从事分子植物病毒学研究，E-mail：yingwang@cau.edu.cn。

中国小麦花叶病毒 CP 和 CRP 蛋白的原核表达、抗血清制备及 RNA2 侵染性克隆构建

孔凡惠[1]，脱建波[1]，魏娇[1]，唐伟[1]，李向东[1*]，迟胜起[2]，田延平[1]，于金凤[1]

([1]山东农业大学植物保护学院植物病理学系，泰安 271018；

[2]青岛农业大学农学与植保学院，青岛 266109)

摘　要：中国小麦花叶病毒（*Chinese wheat mosaic virus*，CWMV）引起的小麦土传花叶病在山东烟台、威海等地危害严重。本研究克隆了 CWMV 烟台分离物的外壳蛋白（Coat Protein，CP）及富含半胱氨酸蛋白（Cysteine-rich protein，CRP）基因，并将其连接到原核表达载体 pEHISTEV，转化大肠杆菌 Rosetta。经 IPTG 诱导，表达出分子量均为 19kDa 的 CP 和 CRP。将二者从凝胶中切下，乳化后免疫新西兰大耳兔 4 次，获得了两种蛋白的多克隆抗体。ELISA 检测表明，CWMV CP 和 CRP 抗血清的效价分别为 1∶4 096 和为 1∶2 048。Western blot 分析证明该抗血清只与感染 CWMV 的小麦有特异性反应，而与健康小麦或感染小麦黄花叶病毒的小麦无反应。利用含 T3 启动子的引物通过 RT-PCR 扩增 CWMV RNA2 全长片段，经 T/A 克隆连接到 pMD18-T，获得质粒 pMD18-T-CWMV-RNA2。该质粒经 Xba I 线性化后，利用 T3 RNA 聚合酶进行体外转录，转录产物摩擦接种本氏烟，15℃培养 3 d 后，利用 Western blot 可从接种叶片中检测到瞬时表达的 CWMV CP 蛋白。

关键词：中国小麦花叶病毒；衣壳蛋白；富含半胱氨酸蛋白；抗血清制备；侵染性克隆

孔凡惠和脱建波为共同第一作者。

* 通讯作者

在我国山西和甘肃地区首次检测到苹果坏死花叶病毒

谢吉鹏[1]，龚卓群[1]，陈冉冉[1]，国立耘[1]，李世访[2]，周涛[1*]

([1]中国农业大学植物病理学系，农业部植物病理学重点实验室，北京 100193；

[2]中国农业科学院植物保护研究所，北京 100193)

摘　要：我国是世界苹果生产第一大国，病毒病普遍发生，严重影响我国苹果生产品质和效益，其中苹果花叶病是我国大部分苹果产区常见的一种病毒病。在山西和甘肃地区分别采集了花叶和无花叶症状的苹果叶片样品，其中山西样品中花叶症状 20 份，无花叶症状 4 份；甘肃样品中花叶症状 10 份，无花叶症状 16 份。所采集的花叶样品症状主要表现为沿叶脉褪绿黄化，叶脉间多为小黄斑或褐色环斑。提取花叶样品和无花叶样品组织总 RNA，通过 RT-PCR 检测，以苹果坏死花叶病毒(*Apple necrotic mosaic virus*，ApNMV)的特异性引物扩增得到 640 bp 的目标片段。在山西和甘肃的 30 份花叶样品中，ApNMV 阳性样品数为 25，阳性率为 83.3%；在 20 份无花叶样品中，ApNMV 阳性样品数为 6，阳性率为 30.0%。

关键词：苹果花叶病；苹果坏死花叶病毒；病毒检测；RT-PCR

基金项目：国家现代苹果产业技术体系(nycytx-08-04-02)。

* **通讯作者：**周涛，副教授，主要从事植物病毒学研究，E-mail：taozhoucau@cau.edu.cn。

云南部分地区苹果样品病毒和类病毒的检测

谢吉鹏[1]，王少杰[1]，龚卓群[1]，陈冉冉[1]，马钧[2]，国立耘[1]，周涛[1*]

([1]中国农业大学植物病理学系，农业部植物病理学重点实验室，北京 100193；

[2]云南省农业科学院，昆明 650231)

摘　要：苹果病毒病已成为严重影响我国苹果产业发展和生产效益的重要因素，苹果潜隐病毒中以苹果褪绿叶斑病毒(*Apple chlorotic leaf spot virus*，ACLSV)、苹果茎沟病毒(*Apple stem grooving virus*，ASGV)和苹果茎痘病毒(*Apple stem pitting virus*，ASPV)分布最广，为害最大，会导致树势衰退，严重影响果实产量和品质。苹果锈果类病毒(*Apple scar skin viroid*，ASSVd)的危害主要表现为果实表面产生锈斑或因着色不均引起红绿相间的花脸状。分别以 ACLSV(794 bp)、ASGV(500 bp)、ASPV(316 bp)和 apscarviroid(220 bp)的特异性检测引物对采集于云南省马龙县和永善县的52份苹果样品进行了病毒检测。结果表明，ACLSV 阳性样品数为 36，阳性率为 69.2%；ASGV 阳性样品数为 37，阳性率为 71.2%；ASPV 阳性样品数为 28，阳性率为 53.8%；ASSVd 阳性样品数为 4，阳性率为 7.69%。其中，ACLSV、ASGV 和 ASPV 三种病毒复合侵染的样品数为 25，复合侵染率为 48.1%。

关键词：苹果潜隐病毒；苹果锈果类病毒；病毒检测；RT-PCR

基金项目：国家现代苹果产业技术体系(nycytx-08-04-02)。

* **通讯作者**：周涛，副教授，主要从事植物病毒学研究，E-mail：taozhoucau@cau.edu.cn。

小麦黄花叶病毒衣壳蛋白的原核表达及抗血清制备

唐伟[1]，程德杰[1]，魏娇[1]，孔凡惠[1]，李向东[1,2*]，于金凤[1,2]
([1]山东农业大学植物保护学院植病系，泰安 271018；
[2]山东省小麦玉米周年高产高效协同创新中心，泰安 271018)

摘　要：通过 RT-PCR 的方法扩增获得小麦黄花叶病毒(*Wheat yellow mosaic virus*，WYMV)的衣壳蛋白(CP)基因，并将其连接原核表达载体 pEHISTEV。将重组质粒转化到大肠杆菌 Rosetta，通过 IPTG 诱导后，可以表达 38 kDa 的融合蛋白。通过切胶回收的方法收集目的蛋白，免疫新西兰长耳兔，制备 WYMV CP 的多克隆抗体。间接 ELISA 测定该抗血清效价为 1∶2 048，Western blotting 分析表明该血清可以与病株中 CP 发生特异性反应，而与健康植株汁液无反应。

关键词：小麦黄花叶病毒；外壳蛋白；原核表达；抗血清；检测

基金项目：农业公益性行业科技项目(201303021)；山东省现代农业产业技术体系。

唐伟和程德杰为共同第一作者。

* **通讯作者**

芜菁花叶病毒 P3 蛋白与拟南芥 AtSWEET1 蛋白的互作研究

张雅琦[1] 祝富祥[1] 王艳[1] 李向东[2] 潘洪玉[1] 刘金亮[1*]

([1]吉林大学植物科学学院,长春 130062;[2]山东农业大学植物保护学院,泰安 271018)

摘　要:芜菁花叶病毒（*Turnip mosaic virus*，TuMV）是马铃薯 Y 病毒科(*Potyviridae*)马铃薯 Y 病毒属(*Potyvirus*)的重要成员。该病毒具有广泛的寄主范围,可侵染至少 156 属的 300 多种植物,尤其对十字花科蔬菜造成严重威胁,同时也是传播最广、破坏性最强的侵染芸薹属植物的病毒,给农业生产带来巨大损失。TuMV 基因组是正义单链 RNA,包含一个大的开放阅读框(ORF),*P*3 蛋白是该病毒编码的一个重要蛋白,相比其他蛋白,P3 蛋白高度变异,对其结构和功能了解较少。目前已知,P3 是一个膜蛋白,在 TuMV 侵染寄主的过程中发挥着一定的作用,可能涉及决定致病性、病毒移动、病毒复制、症状形成和寄主抗性等方面。为了解 TuMV P3 蛋白与寄主植物之间的互作关系,利用 TuMV JCR06 分离物 *p*3 基因 (GenBank 登录号:KP165425) 构建诱饵载体,通过酵母双杂交融合的方法,从拟南芥酵母双杂交 cDNA 文库中筛选到一个与 P3 蛋白互作的糖转运蛋白(AtSWEET1),并用共转化的方法进行了验证两者之间的互作。对 *p*3 基因 C 端进行缺失,利用 *p*3 基因 N 端 663 bp 构建诱饵载体,并与 AtSWEET1 基因猎物载体共转化酵母,进一步验证 AtSWEET1 和 *p*3 基因 N 端 663 bp 同样存在互作。将 *p*3 基因和 AtSWEET1 基因分别连接到植物表达载体 pCG-1301 上,转化农杆菌 EHA105,侵染本生烟进行亚细胞定位分析,结果显示二者均存在于本生烟表皮细胞的细胞膜上。在此基础上,利用双分子荧光互补技术(BiFC)验证 P3 蛋白与 AtSWEET1 蛋白之间的相互作用,结果表明二者在洋葱内表皮细胞内互作,且作用部位位于细胞膜,推测两个蛋白均为膜蛋白,这为研究 TuMV P3 蛋白与 AtSWEET1 蛋白的互作机制及其在 TuMV 致病过程中的功能奠定了理论基础。

关键词:芜菁花叶病毒;P3;糖转运蛋白 AtSWEET1;蛋白互作

基金项目:国家自然科学基金项目(31201485)。

作者简介:张雅琦,女,吉林珲春人,硕士研究生,主要从事分子植物病毒学的研究,E-mail:zhang_ya_qi@163.com。

[*] **通讯作者**:刘金亮,副教授,主要从事植物病毒学与分子植物病理学的研究,Tel:0431-87835707,E-mail: jlliu@jlu.edu.cn。

为害山东芝麻的病毒种类检测

王红艳[1,2],宫慧慧[1],赵鸣[1],孟庆华[1],马惠[1],李向东[2]
([1]山东棉花研究中心,济南 250100;[2]山东农业大学植物保护学院植病系,泰安 271018)

摘　要:芝麻属胡麻科胡麻属,是我国四大油料作物之一。我国芝麻常年种植面积为 80 万 hm^2左右,产量居世界第三。山东具有悠久的芝麻种植历史,全省芝麻种植面积超过 2 万 hm^2。山东省地处黄河下游,气候温和,雨量充沛,利于多种芝麻病害的发生,其中芝麻病毒病是当前限制其产量和质量的重要病害。为了明确山东主要芝麻病毒病害的发生情况,了解不同芝麻品种(品系)中病毒病的侵染情况和发病症状,本研究从 3 个山东芝麻产区采集芝麻样品,同时从育种基地采集不同品种(品系)的芝麻样品,并对这些样品进行 PCR 检测和序列分析。共检测出 3 种 RNA 病毒:黄瓜花叶病毒(*Cucumber mosaic virus*, CMV),烟草花叶病毒(*Tobacco mosaic virus*, TMV),西瓜花叶病毒(*Watermelon mosaic virus*, WMV)。在 25 个病毒阳性样品中,受 2 种以上病毒复合侵染的占多数。受病毒复合侵染的芝麻症状表现复杂,表现为花叶、黄化、皱缩、线叶、果实畸形等,且同一种病毒在不同品种(品系)上表现的症状也稍有差异。

作者简介:王红艳,女,助理研究员,在读博士,主要从事植物病害生物防治、植物病毒学研究,E-mail:sdauwhy@163.com。

为害广东冬种辣椒主要病毒种类的鉴定

汤亚飞[1,2],佘小漫[1],何自福[1,2*],蓝国兵[1]

([1]广东省农业科学院植物保护研究所,广州 510640;

[2]广东省植物保护新技术重点实验室,广州 510640)

摘　要:辣椒是广东主要冬种蔬菜之一,年种植面积达 100 万亩以上。病毒病是为害冬种辣椒的主要病害之一,每年均造成不同程度损失。为了弄清广东冬种辣椒病毒病为害情况及病原病毒种类,2013—2015 年,本团队在茂名和湛江等冬种蔬菜产区进行调查,发现辣椒病毒病发生较严重,田间病株率 10%～60%,严重时达到 100%;田间症状主要表现为花叶、黄化、蕨叶、畸形、脉斑驳等症状,以花叶和脉斑驳为主。利用 sRNA 深度测序技术,对广东省茂名市等冬种辣椒主要产区病样进行了病毒种类鉴定,进一步根据 sRNA 测序结果设计每种病毒的特异引物,进行 RT-PCR 验证。初步结果表明为害广东冬种辣椒的病毒种类多达 12 种,分别为辣椒环斑病毒(*Chilli ringspot virus*)、蚕豆萎蔫病毒(*Broad bean wilt virus*)、灯笼椒内源病毒(*Bell pepper endornavirus*)、黄瓜花叶病毒(*Cucumber mosaic virus*)、辣椒黄脉病毒(*Pepper vein yellows virus*)、红辣椒脉斑驳病毒(*Chilli veinal mottle virus*)、辣椒脉斑驳病毒(*Pepper veinal mottle virus*)、辣椒黄化曲叶病毒(*Pepper yellow leaf curl virus*)、辣椒褪绿病毒(*Capsicum chlorosis virus*)、烟草轻型绿花叶病毒(*Tobacco mild green mosaic virus*)、辣椒潜隐病毒(*Pepper cryptic virus*)、马铃薯 Y 病毒(*Potato virus* Y)。

关键词:冬种辣椒;病毒种类;sRNA 深度测序技术

基金项目:国家公益性行业(农业)科研专项(201303028);广东省科技计划项目(2013B020309003;2014B070706017)。

作者简介:汤亚飞,助理研究员,主要从事植物病毒学研究。

* **通讯作者:**何自福,研究员,主要从事蔬菜病理学研究,E-mail:hezf@gdppri.com。

甜瓜坏死斑点病毒侵染性克隆的构建

吴会杰，古勤生*
（中国农业科学院郑州果树研究所，郑州 450009）

摘　要：甜瓜坏死斑点病毒（*Melon necrotic spot virus*，MNSV）属香石竹斑驳病毒属（*Carmovirus*），病毒粒体为球形，直径约 30 nm，基因组为正义单链 RNA 约 4.3 Kb，5′端不含帽子结构，且 3′端不含 poly(A) 结构，编码 5 个开放阅读框。它主要通过种子、土壤真菌和黄瓜黑头叶甲进行自然传播，另外还可以通过机械摩擦接种进行传播。其寄主范围仅限于葫芦科的一些植物，如西瓜、南瓜、葫芦、黄瓜和甜瓜等。

为了构建 MNSV 的侵染性克隆，本研究将甜瓜坏死斑点病毒的全基因组融合到植物表达载体质粒 pXT1（南京农业大学陶小荣老师赠）中，随后利用农杆菌注射接种的方法接种甜瓜两叶一心的幼苗，接种后把甜瓜苗保持在黑暗中 10 h 后，在 20℃ 8 h，24℃ 16 h 的条件下培养。

接种后每天观察发病情况。接种第 4 天后，接种的子叶开始出现局部的坏死斑点症状，至第 7 天后时，子叶的枯斑开始连片出现，真叶也出现明显的坏死斑点症状。注射接种子叶第 7 天后植株相继发病，接种叶出现大量枯斑，枯斑连片呈现严重枯死，出现的第一个真叶产生典型的枯斑，同时，经 ELISA 和 RT-PCR 检测显示，接种 MNSV 侵染性克隆的大部分植株表现阳性，发病率达 95%（发病 37/39）。随后将发病的植株经摩擦接种回接到健康的甜瓜植株上，接种第 7 天后植株均发病，说明构建的侵染性克隆能够正常传播。由此，MNSV 侵染性克隆构建成功。

基金项目：国家现代产业技术体系（CARS-26-13）；中国农业科学院创新团队（CAAS- ASTIP-2015-ZFRI）；国家自然基金项目（31071811）。

作者简介：吴会杰，女，主要从事植物病理学研究，Tel：0371-65330956，E-mail：wuhuijie@caas.cn。

* **通讯作者**：古勤生，男，主要从事植物病理学研究，Tel：0371-65330997，E-mail：guqinsheng@caas.cn。

双生病毒抑制茉莉酸和乙烯抗虫通路与烟粉虱形成互惠共生关系

赵平芝[1]，姚香梅[1]，孙艳伟[1]，马永焕[1]，王晓伟[2]，刘树生[2]，周雪平[3]，叶健[1*]

([1]中国科学院微生物研究所，植物基因组学国家重点实验室；
[2]浙江大学昆虫科学研究所；[3]中国农科院植物保护研究所)

摘　要：近80%的植物病毒都是经昆虫介体来实现植物—植物间的传播，病毒—传毒昆虫—寄主植物三者形成了最简单的多界互作生态系统。我们选取双生病毒—植物—烟粉虱为模式系统来研究复杂生物间互作信号的产生、加工、识别和破译的分子机制。双生病毒引起的病害在最近20年间由局部发生的小病害衍变成目前全球性的最重要的植物病毒病害之一，同时双生病毒的介体昆虫烟粉虱在此期间发展成为重要的全球性害虫。生态学研究结果发现双生病毒可以与烟粉虱形成互惠共生的关系(Ruan et al.，2012)，但是其互惠共生的分子机制却不清楚。我们前期的研究成果揭示了双生病毒可以通过抑制植物转录因子MYC2来抑制茉莉酸介导的抗虫性，包括昆虫嗅觉识别的主要信号分子-萜烯类化合物挥发物和芥子油苷等(Li et al.，2014)。我们最新研究还发现植物乙烯抗性信号途径也被双生病毒所抑制。通过遗传和分子生物学分析，我们鉴定了一个新的转录因子参与了乙烯介导的抗虫途径，该转录因子同MYC2一起共同通过直接调控下游基因的表达，对植物抗虫信号网络进行调控；而双生病毒通过同时抑制这两种抗虫信号途径，与烟粉虱形成了依赖于植物的间接互惠共生关系。我们的研究为通过提高作物的抗虫性来防治病毒病害提供了重要理论依据，为发展高效昆虫趋避剂提供了化合物前体。

* **通讯作者**：叶健，E-mail：jianye@im.ac.cn。

双链 RNA 技术在植物病毒病监测中的应用

张俊[1]，胡荣[1]，李政[1]，方守国[1]，郭灵芳[2]，李凡[3]，章松柏[1*]
([1]主要粮食作物产业化湖北省协同创新中心，荆州 434025；
[2]长江大学工程技术学院，荆州 434020；[3]云南农业大学植物保护学院，昆明 650201)

摘　要：正常的植物体内很难检测到双链 RNA (double-stranded RNA, dsRNA)的存在，但是植物遭受病原生物入侵后，往往产生大量的 dsRNA，有些是植物应对病原物入侵而产生的小分子 dsRNA，有些是 RNA 病毒入侵寄主而产生的复制中间体(replicative intermediate, RI)或病毒本身的基因组核酸就是 dsRNA。鉴于 90%的植物病毒病是 RNA 病毒引起的情况，探讨了 dsRNA 技术(dsRNA 提取和克隆技术)在植物病毒病监测中的应用。具体应用如下：①dsRNA 病毒病的监测，可选择双链 RNA 快速小量提取法，它能够从 0.5～0.8 g 的植物组织中快速提取 dsRNA，借助 dsRNA 病毒特异的基因组图谱完成对样品的鉴定，在水稻病毒病的监测中得到了应用；②单链 RNA 病毒病的监测，可选择双链 RNA 快速大量提取法，即富集复制中间体的方法，从 5～10 g 植物组织中富集病毒的 dsRNA，然后通过克隆和序列分析完成鉴定，该方法在蔬菜和花卉病毒病中得到了应用；③新病毒的挖掘，借助 dsRNA 技术，发现了 6 种新的病毒，其中 3 种病毒的基因组已登录 Genbank，即 *Brassica compestris chrysovirus* 1 (KP782029、KP78203、KP782031)、*Panax notoginseng virus A*(KT388111)、*Erysiphe cichoracearum endornavirus*(KT388110)。因此，双链 RNA 技术可以作为植物病毒病监测的一种重要手段。

关键词：双链 RNA 技术；dsRNA 病毒；复制中间体

基金项目：农业部公益性行业(农业)科研专项(201303028)；国家自然科学基金青年基金项目(31301638)。

作者简介：张俊(1992—　)，男，湖北省松滋市人，硕士生，研究方向为植物病毒监测，Tel：13872278636，E-mail：191172885@qq.com。

*__通讯作者：__章松柏(1978—　)，男，湖北省黄梅县人，博士，副教授，研究方向为植物病毒的监测和分子病毒学，Tel：18972361635，E-mail：yangtze2008@126.com。

农杆菌介导的黄瓜绿斑驳花叶病毒侵染性克隆的构建及其相关的突变

刘莉铭，彭斌，古勤生*

（中国农业科学院郑州果树研究所，河南省果树瓜类生物学重点实验室，郑州 450009）

摘　要：黄瓜绿斑驳花叶病毒（*Cucumber green mottle mosaic virus*，CGMMV）是葫芦科上的重要病毒之一，严重为害葫芦科作物，威胁产业发展。近几年我国嫁接西瓜由于带毒砧木种子的滥用，该病毒曾在不同省份暴发。

本实验室根据已有 CGMMV-hn 分离物的序列，将其基因组分成两段扩增，依次插入植物表达载体 pXT1 中，从而获得 CGMMV 全长克隆 pXT1-CGMMV。为了验证该克隆的侵染性，采用农杆菌注射法接种本生烟、西瓜、甜瓜和瓠瓜，2～3 周后植株相继产生典型的花叶、斑驳症状，经 ELISA 和 RT-PCR 检测，发病植株均为阳性。之后将发病的叶片经摩擦接种回接到健康的西瓜、甜瓜、本生烟植株上，植株均能发病，说明构建的侵染性克隆具有强致病性，可以用于突变体的筛选、病毒与寄主互作等相关的研究。

本实验室应用 pXT1-CGMMV，对复制酶区域进行了突变检测，以期了解该区域与病毒致病性的关系。利用定点突变技术共获得 49 个突变体，ELISA 检测显示 21 个突变体为阳性，但部分突变体在植株体内病毒含量较低或不造成明显症状或使植株发病明显延迟。症状形成的分子决定簇正进一步分析中。

关键词：黄瓜绿斑驳花叶病毒；侵染性克隆；突变

基金项目：国家西甜瓜产业技术体系（CARS-26-13）；中国农业科学院创新工程（CAAS-ASTIP-2015-ZFRI-08）；国家自然科学基金项目（31572147）。

作者简介：刘莉铭，博士研究生，植物病理学专业。

* **通讯作者**：古勤生，研究员，E-mail：guqsh@126.com。

梨带病毒和无病毒植株生理和生化特性比较

陈婕[1]，汤慧慧[1]，李庆余[3]，文立红[1]，洪霓[1,2*]，王国平[1,2]

([1]华中农业大学植物科技学院；[2]华中农业大学，农业微生物学国家重点实验室，武汉 430070；[3]山东省烟台市农业科学院)

摘　要：苹果茎沟病毒(*Apple stem grooving virus*，ASGV)是苹果和梨树上发生普遍的病毒，该病毒还可危害柑橘和猕猴桃，对这些果树造成不同程度的危害。培育和栽培无病毒种质是防止该病毒危害的有效途径。为明确该病毒对梨树生长和生理生化特性影响，本研究以无病毒和带有 ASGV 的沙梨"金水 2 号"离体植株和田间嫁接一年生植株为研究材料，分析比较了在生根培养基(1/2MS＋2.5 mg/L IBA＋1.5 mg/L NAA)上二者的生根效果，结果显示，在该培养基上无病毒"金水 2 号"离体植株生根率可达 90%，而带 ASGV 的离体植株在该生根培养基中完全不能生根，表明该病毒对"金水 2 号"离体植株的生根有严重的抑制作用。测定了无病毒和带有 ASGV 的沙梨"金水 2 号"田间嫁接一年生植株叶片的叶绿素含量及苯丙氨酸解氨酶(PAL)、谷氨酰胺合成酶(GS)、超氧化物歧化酶(SOD)和过氧化物酶(POD)活性，结果显示，带病毒植株的 PAL 酶和 SOD 酶活性显著高于无毒植株，GS 酶活性显著低于无毒植株，而叶绿素含量和 POD 酶活性没有显著差异，该结果初步表明 ASGV 病毒侵染沙梨"金水 2 号"植株后会提高植株体内与抗逆性相关的酶活性，而降低氮代谢相关的 GS 酶活性。

基金项目：国家农业部公益性行业计划(201203076-03)；梨现代农业技术产业体系(nycytx-29-08)。

* **通讯作者**：洪霓，E-mail：whni@mail.hzau.edu.cn。

侵染猕猴桃的番茄斑萎病毒属病毒鉴定

王雁翔，杨作坤，朱晨熹，郑亚洲，洪霓*，王国平
（华中农业大学植物科学技术学院，武汉 430070）

摘　要：近些年，我国猕猴桃种植面积和产量均呈快速上升趋势，同时病毒病危害也正在加剧。目前世界上已报道关于猕猴桃病毒病有 13 种，前期本研究组报道了侵染我国猕猴桃的 3 种病毒，即猕猴桃病毒 A（*Actinidia virus A*，AcVA），猕猴桃病毒 B（*Actinidia virus B*，AcVB）和柑橘叶斑驳病毒（*Citrus leaf blotch virus*，CLBV）。为深入了解我国发生的猕猴桃病毒种类，本研究采用小 RNA（small RNA，sRNA）高通量测序技术，从采集于贵州的猕猴样品中检测到一种番茄斑萎病毒属病毒。通过对 sRNA 序列拼接后获得 contigs 在 GenBank（http://www.ncbi.nlm.nih.gov/）上进行 BLASTX 和 BLASTP 比对，搜索病毒相关序列，共有 464 条 contig 的氨基酸序列与番茄斑萎病毒属病毒编码的蛋白序列相似性较高，根据所获序列设计特异性引物和末端序列扩增，获得该病毒全基因组序列。3 条 RNA（L，M，S）大小分别为 8 909、4 772 和 2 991 nt，其中 L RNA 和 M RNA 与已知番茄斑萎病毒属病毒核苷酸序列相似性均小于 80%，而 S RNA 与我国报道的番茄坏死斑点病毒（*Tomato necrotic spot tospovirus*，TNSV）核苷酸序列相似性达 97%。这是首次从猕猴桃上鉴定到番茄斑萎病毒属病毒。

基金项目：国家农业部公益性行业计划（201203076-03）。

* **通讯作者**：洪霓，E-mail：whni@mail.hzau.edu.cn。

Xanthomonas campestris pv. *raphani* 756C中Ⅵ型分泌蛋白生物信息学分析

韩长志

（西南林业大学林学院，云南省森林灾害预警与控制重点实验室，昆明 650224）

摘　要：病原细菌与植物相互作用过程中，病原细菌为了更好地侵染植物，可通过分泌系统将分泌蛋白输入到寄主组织中，这些分泌蛋白可作为效应分子来与植物中的防卫反应相关分子发生作用，从而得以在植物中实现定殖、扩展等过程。目前，在细菌中已经发现有7种分泌系统，包括Ⅰ型分泌系统（Type Ⅰ secretion system，T1SS）到Ⅶ型分泌系统（Type Ⅶ secretion system，T7SS）。不同学者对植物细菌T1SS、Ⅱ型分泌系统，耶尔森菌（*Yersinia*）、绿脓杆菌（*Pseudomonas aeruginosa*）、志贺菌（*Shigella*）的Ⅲ型分泌系统，以及对Ⅳ型分泌系统、Ⅴ型分泌系统从其结构、功能、调控等方面进行了大量研究。Ⅵ型分泌系统（Type Ⅵ secretion system，T6SS）是2006年在人类致病细菌——霍乱弧菌（Vibrio cholerae）中发现的一种新的分泌系统，广泛存在于革兰氏阴性细菌中。国内外学者从分泌系统结构、生物学功能与调控等方面对植物细菌T6SS进行了大量研究，已经初步明确植物青枯菌（*Ralstonia solanacearum*）、伤寒沙门菌（*Salmonella enterica serovar* Typhi）等革兰氏阴性细菌T6SS基因的功能。根据T6SS各组分所具有的功能，可将其分为结构蛋白、形成跨膜通道结构的转位蛋白、分泌蛋白以及对分泌系统起辅助作用的蛋白等，然而对于分泌蛋白在致病细菌的生理、致病等过程中所具有的功能尚不清楚。同样，对于可以引起茄科蔬菜和十字花科蔬菜叶斑病、给全世界的经济产生重大损失的黄单胞菌（*Xanthomonas campestris* pv. *raphani* 756C，Xcr）T6SS蛋白功能的研究也鲜有报道。本研究通过对已完成基因组测序和蛋白功能注释的Xcr进行Ⅵ型分泌蛋白（Type six secretion protein，Tss）的搜索，通过保守结构域分析、疏水性分析、理化性质分析、二级结构和三级结构预测、信号肽分析以及跨膜结构域分析、亚细胞定位、motif分析等生物信息学方法，以期明确该菌中Tss的理化性质、保守结构域特征以及亚细胞定位情况等，同时，基于上述发现的Tss氨基酸序列，在美国国家生物信息中心（NCBI）进行Blastp同源序列搜索，通过遗传关系分析，以期为进一步开展同属于黄单胞菌属但其基因组序列尚未公布的核桃细菌性黑斑病菌的研究提供重要的理论指导。

关键词：*Xanthomonas campestris* pv. *raphani* 756C；Ⅵ型分泌系统；Ⅵ型分泌蛋白；黄单胞菌属

基金项目：云南省优势特色重点学科生物学一级学科建设项目（No. 50097505）；云南省高校林下生物资源保护及利用科技创新团队（No. 2014015）；国家自然科学基金青年科学基金项目（No. 31200488）。

作者简介：韩长志（1981—　），男，博士，讲师，河北省石家庄市人，主要从事经济林木病害生物防治与真菌分子生物学研究，E-mail：hanchangzhi2010@163.com。

河北省甘薯茎腐病的发生及其病原鉴定

高波,王容燕,马娟,李秀花,陈书龙*
(河北省农林科学院植物保护研究所,河北省农业有害生物综合防治工程技术研究中心,
农业部华北北部作物有害生物综合治理重点实验室,保定 071000)

摘　要:甘薯茎腐病是近几年在我国发现的一种新的细菌性病害,目前该病已在我国的福建、广东、江西、广西、海南、河南、重庆、江苏等地发生。2013—2015 年我们在河北文安县甘薯育苗期和收获期的病害调查中,发现疑似甘薯茎腐病的病害,育苗期主要症状表现为种薯呈水浸状腐烂,颜色变黑,伴有恶臭味,烂薯无法出苗或出苗后发生薯苗整株萎蔫腐烂;收获期主要症状表现为甘薯植株茎基部腐烂变褐,地下薯块呈水浸状腐烂,掰开后呈"烤红薯"状,田间发病率在 5%左右($n=50\times3$),受害植株基本无产量,给当地的甘薯生产造成了严重的影响。经对疑似病株病原菌的分离纯化以及柯赫氏法则验证,结果显示细菌菌株 WA1301(来自收获期)和 WA1503(来自育苗期)均可以导致甘薯薯块和茎蔓发生水浸状腐烂,并与田间病株症状一致,说明两株菌株为病原菌。通过形态观察,病原菌 WA1301 和 WA1503 菌体形态基本一致,均为短杆状,大小为(0.5～0.8)μm ×(1.0～2.7)μm,革兰氏染色为阴性,在 LB 平板培养基上 30℃培养 24 h 后,菌落呈浅黄色,边缘整齐。对该两株病原菌 16S rDNA 的克隆和序列分析结果显示,该两株菌间相似性达 100%,与 Huang 等(2010)报道的 *Dickeya dadantii*(又名 *Erwinia chrysanthemi*)菌株 H12(GU252371)和 09-1(HM222417)的相似性均达 99%,另外系统发育分析结果显示,其均与 *D. dadantii* 聚为一组,进一步证明了病原菌 WA1301 和 WA1503 为同种细菌,且均属于达旦提狄克氏菌(*D. dadantii*)。通过回接病株症状,病菌形态学观察以及分子鉴定,最终确定该病害为甘薯茎腐病,病原菌为达旦提狄克氏菌。这是该病害首次在河北省被发现。达旦提狄克氏菌致病力较强,在甘薯的收获期和育苗期均能侵染为害,且一旦发病将对甘薯造成毁灭性的严重后果。因此,甘薯茎腐病的发生应该引起北方甘薯种植区的足够重视,从选育抗病品种、培育无病壮苗、田间排水排涝、硫酸链霉素浸秧喷雾等入手,做好综合防治工作。

关键词:甘薯茎腐病;细菌性病害;达旦提狄克氏菌

基金项目:国家甘薯产业技术体系(CARS-11-B-08);河北省自然科学基金(C2015301043);河北省财政基本科研业务费(494-0401-JBN-6440)。

作者简介:高波(1985—　),男,助理研究员,主要从事甘薯病害研究,E-mail:gaobo89@163.com。

* **通讯作者:**陈书龙,研究员,主要从事植物寄生线虫学研究,E-mail:chenshulong@gmail.com。

辣椒溶杆菌(*Lysobacter capsici*)X2-3 抗菌作用特点及全基因组序列分析

衣静莉,王静,刘朝霞,张莉,刘爱新*
(山东农业大学植物保护学院,泰安 271018)

摘　要:溶杆菌(*Lysobacter*)多分布于土壤和水体,是植物根围常见的 PGPR 类型之一。本实验室从小麦根际土壤中分离到菌株 X2-3,经 16S RNA 基因序列分析,确定其为辣椒溶杆菌(*Lysobacter capsici*)。抑菌测定发现,菌株 X2-3 对 *Rhizoctonia cerealis*、*Rhizoctonia solani*、*Bipolaris sorokiniana*、*Fusarium oxysporum* f. sp. *cucumerinum*、*Verticillium dahliae*、*Colletotrichum gloeosporioides*、*Pythium myriotylum*、*Botrytis cinerea*、*Botryosphaeria ribis*、*Valsa mali*、*Phytophthora parasitica* var. *nicotianae* 等多种植物病原真菌和卵菌及 *Bacillus* sp. 等革兰氏阳性细菌具有明显的拮抗作用,但对 *Ralstonia solanacearum* 没有作用。为了解 X2-3 中的抗菌物质及其合成基因,在 Illumina MiSeq 及 Illumina HiSeq 2 500 测序平台对 X2-3 基因组进行序列测定,在此基础上,使用 SOAP denovo 软件对所测序列的信息进行组装,并采用 GeneMarkS 软件进行基因预测分析,发现,X2-3 的基因组总长为 6 126 365 bp,组装后产生 3 个 scaffolds,分为 13 个 contigs,*N*50 长度为 699 639 bp。最长和最短的 contig 分别为 1 781 518 bp 和 6 232 bp。对所预测的基因在 NCBI 中进行 Blast 分析发现,X2-3 全基因组中含有 5 个非核糖体多肽合成酶基因簇,它们可能参与抗菌物质的生物合成,但这些基因的功能有待进一步实验验证。

关键词:辣椒溶杆菌(*Lysobacter capsici*);抗菌作用;全基因组序列分析;非核糖体肽合成酶

* 通讯作者

无致病力青枯雷尔氏菌突变菌株的构建及其防效评价

郑雪芳，刘波，朱育菁，陈德局
（福建省农业科学院农业生物资源研究所，福州 350003）

摘　要：由致病性青枯雷尔氏菌（*Ralstonia solanacearum*）引起的番茄青枯病是一种毁灭性的土传病害，一旦发病就难以控制，严重制约着番茄产业的发展和经济效益的提高，目前尚未出现有效的防治措施。利用青枯雷尔氏菌的弱/无致病力菌株防治作物青枯病具有一定生防潜力。本研究通过分离筛选自然弱毒株、^{60}Co 辐射诱变和 EZ-Tn5 插入诱变，分别获得 12、12 和 40 株青枯雷尔氏菌无致病力突变菌；突变后菌株的菌落和菌体形态与出发菌株均有明显的差别。致病性检测试验表明，突变菌株接种番茄盆栽苗 15 d 后均未发病，证实均为无致病力青枯雷尔氏菌。进一步对番茄青枯病的防治试验表明，从番茄青枯病发病田块分离的无致病力突变菌株 FJAT-1458 的防治效果最好，防效达 100%。利用高效液相色谱对无致病力突变菌株进行表征，64 株无致病力突变菌的色谱行为可划分为 3 种类型（单峰型、双峰型和三峰型），其中单峰型 8 株，双峰型 35 株，三峰型 21 株；构建“色谱效价指数”（chromatography titer index，CTI），CTI>90%的无致病力突变菌有 24 株，CTI＝100%的有自然弱毒株 FJAT-1458、3 株 ^{60}Co 辐射突变株和 4 株 EZ-Tn5 插入突变株。分析无致病力突变菌的 CTI 与其对番茄青枯病防效的相互关系，结果表明二者呈极显著相关（$P>0.01$）；定殖试验表明，FJAT-1458 可定殖于番茄植株根系土壤、根部和茎部，定殖数量均表现为“先增后减”的趋势；并且接种浓度越大、苗龄越小，其定殖数量越大。对番茄青枯病的防治效果表明，预接种 FJAT-1458 的浓度越高，防效越好。

关键词：青枯雷尔氏菌；突变菌株；防效评价

沙姜青枯病菌 YC45 菌株 *hrpB* 突变株的构建

佘小漫[1,2]，何自福[1*]，汤亚飞[1]，蓝国兵[1]

([1]广东省农业科学院植物保护研究所，广州 510640；

[2]广东省植物保护新技术重点实验室，广州 510640)

摘　要：茄科青枯菌(*Ralstonia solanacearum*)是最具毁灭性的植物病原细菌之一，该病原菌可以侵染 54 个科 450 多种植物引起青枯病。不同来源的青枯菌对不同寄主植物的致病性不同。青枯菌致病性和寄主特异性主要通过青枯菌基因调控网络来实现的，其中Ⅲ型分泌系统及其分泌的效应蛋白与青枯菌致病性密切相关。*hrpB* 基因不仅调控Ⅲ型分泌系统组成基因以及多种效应子基因的表达，而且控制依赖于Ⅲ分泌系统输出途径相关基因的表达。因此，*hrpB* 对青枯菌的致病性有重要影响。茄科青枯菌 4 号生理小种引起姜瘟和沙姜瘟是广东及华南地区重要病害，每年均有不同程度的发生与为害。本研究以茄科青枯菌沙姜菌株 YC45(Race 4，Biovar IV)为对象，通过 PCR 扩增 YC45 *hrpB* 基因 ORF 上下游 500 bp 序列，以庆大霉素抗性基因为标记基因，将 *hrpB* 基因上游和下游片段，以及庆大霉素抗性基因，按照上游-庆大霉素-下游的顺序依次连接到 pK18mobsacB 载体中，通过酶切鉴定获得重组自杀质粒 pK18-*hrpB*-Gm 质粒。将重组自杀质粒 pK18-*hrpB*-Gm 质粒电击转化 YC45 菌株，通过三步筛选法最终获得突变菌株 YC45-Δ*hrpB*-Gm，为下一步研究沙姜青枯病菌 YC45 的致病性、效应子及寄主特异性等奠定了基础。

关键词：茄科青枯菌；沙姜菌株；*hrpB*

基金项目：广东省科技计划项目(2014B070706017)；广东省农业科学院院长基金(201513)。

作者简介：佘小漫(1981—)，女，副研究员，硕士，主要从事植物病原细菌研究。

* **通讯作者：**何自福(1966—)，男，研究员，博士，主要从事蔬菜病理学研究，E-mail：hezf@gdppri.com。

葡萄酸腐病相关细菌的分离鉴定及其拮抗菌作用机理

王超男[1]，任争光[1]，李兴红[2]，李红[1]，胡盼[1]，魏艳敏[1*]

([1]北京农学院植物科学技术学院农业应用新技术北京市重点实验室，北京 102206；

[2]北京市农林科学院植保环保所，北京 100097)

摘　要：葡萄酸腐病近年来已经成为葡萄生长中后期的重要病害之一，造成葡萄果穗大量腐烂，严重影响葡萄的品质与产量，制约了葡萄产业的发展。针对上述情况，本研究采用组织分离法对来自延庆的 5 个葡萄酸腐病病果样本进行了相关微生物的分离，共分离出 12 株相关细菌。为了明确这些细菌的致病性，实验采用室内离体果穗人工接种法进行了致病性测定，结果发现其中的两株细菌 sfB-18 和 sfyg3-2 对葡萄果穗具有较强致病性，接种后的葡萄果穗发病率均达到 100%。采用细菌的 16S rDNA 通用引物(27f/1492r)进行 PCR 扩增，并在 NCBI 上对扩增片段进行比对，菌株 sfB-18 被鉴定为醋酸杆菌(*Acetobacter malorum*)，sfyg3-2 为葡萄糖杆菌(*Gluconobacter cerinus*)。实验以菌株 sfB-18 和 sfyg3-2 为靶标菌筛选对葡萄酸腐病有拮抗活性的生防菌，室内抑菌实验发现，本实验室保存的生防菌株解淀粉芽孢杆菌(*Bacillus amyloliquefaciens*)BJ-6 对这两个菌株具有较强的抑菌活性，抑菌带宽度分别为 4 mm 和 5 mm。另外，实验根据已报道芽孢杆菌产生的 4 种环脂肽 Fengycin、Surfactin、Iturin 和 Bacillomycin 的基因(*srfD*、*ituC*、*bamD*、*ppsE*)保守序列设计引物，对菌株 BJ-6(*B. amyloliquefaciens*)进行 PCR 扩增检测，结果显示菌株 BJ-6 能扩增出 4 条相应的目的片段，经测序比对分析验证，推测其可能产生上述 4 种环脂肽，且这些环脂肽在对葡萄酸腐病原细菌的拮抗作用中起到关键作用，明确该芽孢杆菌的抑菌作用机理。试验结果将为葡萄酸腐病的生物防治提供理论依据，为进一步研究生防菌中的不同抗生素的作用奠定基础。

关键词：葡萄酸腐病；分离；鉴定；拮抗菌；抑菌机制

基金项目：北京市自然科学基金(6132008)；现代农业产业技术体系建设专项资金资助(CARS-30-bc-2)。

作者简介：王超男(1991—　)，女，在读硕士，主要从事植物病害防治研究，Tel：010-80794280，E-mail：651168651@qq.com。

* **通讯作者：**魏艳敏(1963—　)，女，教授，主要从事植物病害防治研究，Tel：010-80794280，E-mail：yanminwei@139.com。

陕西省猕猴桃细菌性溃疡病菌群体分子特征与致病力差异分析

赵志博，高小宁，黄丽丽*，秦虎强，康振生

（西北农林科技大学植物保护学院/旱区作物逆境生物学国家重点实验室，杨凌 712100）

摘　要：由 *Pseudomonas syringae* pv. *actinidiae*（PSA）引起的猕猴桃细菌性溃疡病是猕猴桃生产上的毁灭性病害，其发病严重、流行迅速、防治困难，已成为世界性重大细菌病害。考虑到猕猴桃栽培历史较短、病菌出现较晚（1984 年）与再次暴发（2008 年）等特点，对该病害体系中病菌群体结构进行分析，不仅有助于揭示病菌起源与进化机制，而且可指导防治策略的制定、预防病害再次暴发。目前，通过指纹图谱、多位点序列分析（MLSA）及全基因组分析，发现 PSA 具有三个流行群体：PSA-Japan、PSA-Korea 以及 PSA-V（global-outbreak）。然而，作为猕猴桃起源地的中国，其 PSA 群体特征尚不明确。本研究中，我们通过分子标记和致病力差异分析对陕西省 PSA 群体结构进行了研究。MLSA（*rpoD*、*acnB*、*gltA*、*pgi* 和 *gyrB*）结果表明，20 株不同来源（寄主品种、分离部位、时间和地域）的陕西 PSA 菌株均属于 PSA-V 群体。为了进一步分析不同菌株间的分子差异，使用三型效应蛋白基因、毒素基因等 13 个高度分化的基因片段对 50 株陕西 PSA 菌株进行了 UPMGA 分析，结果表明，这些菌株可聚为 3 个分支，分别有 15、18 和 17 株；且其分化特征与菌株来源无关。将 20 株陕西 PSA 菌株分别接种到红阳和海沃德猕猴桃枝条，结果表明，陕西省 PSA 菌株之间存在明显的致病力差异，其差异在两个品种上表现一致，且与 3 个 UPMGA 分支无关。尤其，菌株 M227 致病力几乎完全丧失。通过对同一 UPMGA 分支的强致病力菌株 M401 和弱致病力菌株 M227 进行比较基因组学分析，结果发现，两者仅有 20 个基因存在差异。这些差异基因可能与 PSA 致病性密切相关。本研究从分子特征和致病力方面明确了陕西省 PSA 菌株的群体结构特征，并发现了一株弱致病力菌株，对其进行了全基因组测序，为后续致病机制的深入研究提供了基础。

关键词：MLSA；*Pseudomonas syringae* pv. *actinidiae*；kiwifruit；比较基因组

基金项目：中央高校基本科研基金（No. QN2013007）；高等学校学科创新引智计划（No. B07049）；陕西省科技统筹创新工程计划（No. 2012KTJD03-02）。

作者简介：赵志博，博士研究生，主要从事有害生物生态调控研究，E-mail：zhaozhibol@nwsuaf.edu.cn。

* **通讯作者：**黄丽丽，教授，博士生导师，主要从事植物病害综合治理研究，E-mail：huanglili@nwsuaf.edu.cn。

雌根结线虫抑制其寄主免疫研究进展

贾宁，祝乐天，陈晨，孙思，陈芳妮，廖美德，赵利锋，王新荣*
（华南农业大学农学院，广州 510642）

Reviews on the suppression of host immunity by female root-knot nematodes

Ning Jia, Letian Zhu, Chen Chen, Si Sun, Fangni Chen, Meide Liao, Lifeng Zhao, Xinrong Wang*
(College of Agriculture, South China Agricultural University, Guangzhou 510642, China)

摘　要：根结线虫病的危害来自于重复侵染，雌根结线虫产卵是完成根结线虫重复侵染的必要条件。由于雌根结线虫寄生在植物体内，防治雌根结线虫一直是难题。根据植物免疫学理论，雌根结线虫必须持续抑制寄主免疫反应，才能成功寄生。抑制了雌根结线虫发育，就可以抑制其产卵，从而达到防治根结线虫目的。根结线虫的生活史由卵、2龄幼虫、3～4龄幼虫和成虫（雌虫和雄虫）组成。2龄幼虫是活动的线性线虫，是侵染态。侵染24 h后就可以诱导寄主产生巨型细胞，2龄幼虫固定在寄主体内，从寄主巨型细胞内不断吸取营养。在25℃的条件下，2龄幼虫侵入寄主体内约13 d后，蜕皮变成3～4龄幼虫；侵入寄主体内约20 d后，蜕皮变成成虫（雌虫和雄虫），雄虫游离出寄主，雌虫继续发育，约28 d开始产卵。因此，2龄幼虫与寄主建立寄生关系仅需要很短时间。此后2龄幼虫膨大到发育成成虫需要20 d，处于此阶段线虫体积小，难以操作。然后雌虫发育8 d后开始产卵。目前研究主要集中在侵染态的2龄幼虫诱导寄主巨型细胞产生机理研究，寄主巨型细胞内转录组的变化，以及调控与寄主细胞发育周期有关的基因及信号通路研究，以揭示2龄根结线虫诱导寄主巨型细胞产生的机理。至于侵染后的2龄幼虫和3～4龄幼虫如何维持其与寄主的寄生关系，因此时的根结线虫体积小，难以进行离开寄主之后的实验操作，鲜有研究结果发表。雌根结线虫由于体积较大，在寄主体内时间长于8 d，并已经建立了雌根结线虫死活鉴定技术，雌根结线虫食道腺蛋白提取技术，雌根结线虫特定基因原位杂交技术和靶标基因沉默技术，因此可以以雌根结线虫为突破口，研究雌根结线虫抑制寄主免疫的效应蛋白及其信号通路，进而借助分子生物学技术，培育抗线虫植物材料。另外，在长期的自然进化过程中，植物抗逆的分子机理具有保守性。众所周知，在植物根结线虫与寄主的分子互作过程中，存在一些技术难点，主要是侵入到寄主体内的幼龄根结线虫以个为单位，与寄主互作，但因根结线虫幼虫体积小难以进行分子基因操作。已有的研究结果表明，根结线虫的分子致病机理与其他植物病原也有相似之处。因此雌根结线虫抑制寄主免疫的机理是根结线

基金项目：国家自然科学基金（30771409 和 31171825）；国家留学基金（2014）。

作者简介：贾宁，硕士研究生。

* **通讯作者**：王新荣，教授，主要从事植物线虫学研究，E-mail：xinrongw@scau. edu. cn。

虫抑制寄主免疫的必要组成部分,也可能首先获得突破性研究结果。因此该研究提出了新思路,研究结果将有创新性的科学意义和有较为突出的实践意义。

关键词: 雌根结线虫;抑制;免疫;新思路

黑龙江省大庆和安达地区大豆胞囊线虫生理分化研究

陈井生[1]，李泽宇[1*]，李肖白[1]，田中艳[1]，侯丽[1]，段玉玺[2]，陈立杰[2]
([1]黑龙江省农业科学院大庆分院，大庆 163316；
[2]沈阳农业大学植物保护学院北方线虫研究所，沈阳 1108663)

摘　要：大豆胞囊线虫(Soybean cyst nematode, SCN)是大豆生产上的毁灭性病害，世界上大部分国家的商业化大豆品种都是抗线虫品种。目前，国际上通常用生理小种和HG专化型划分不同大豆胞囊线虫群体对大豆的致病力。大庆和安达试验基地是大豆胞囊线虫的天然病圃，是抗性鉴定的基地，由于长期连作的选择压力在致病性分化的可能。本文采用鉴别寄主法对大庆和安达大豆胞囊线虫天然抗线育种病圃进行生理小种和HG专化型鉴定，生理小种的鉴定参照Riggs和Schmitt的鉴别模式；HG类型的鉴定参照Niblack(2002)提出的鉴别模式。鉴定结果表明：安达地区大豆胞囊线虫优势小种为毒力较强的14号生理小种，HG类型为1.3.4.7；大庆地区为3号生理小种，HG类型为7。本文明确了大庆和安达地区大豆胞囊线虫的毒力类型，为抗线虫大豆育种和线虫防治工作提供理论基础。

关键词：大豆；大豆胞囊线虫；生理小种；HG专化型

Physiological differentiationof soybean cyst nematode in Daqing and Anda area of Heilongjiang province

Jingsheng Chen[1], Zeyu Li[1], Xiaobai Li[1], Zhongyan Tian[1], Li Hou[1], Yuxi Duan[2], Lijie Chen[2]
([1]Daqing Branch, Chinese Academy of Science of Heilongjiang Province, Daqing 163316;
[2]Nematology Institute of Northern China, Department of Plant Protection, Shenyang Agricultural University, Shenyang 110866)

Abstract: The soybean cyst nematode (*Heterodera glycines*) is a important constraint to soybean production in the world. Most commercial soybean varieties are resistant nematode varieties in the world. Knowledge of the virulence phenotypes of soybean cyst nematode is crucial in choosing appropriate sources for breeding resistant varieties and managing the nematode. There are two schemes for characterizing ability of a SCN

基金项目：现代农业产业技术体系(CARS-004-CES 07)；农业部公益性行业科研专项(200903040-03)。

作者简介：陈井生(1982—　)，男，副研究员，博士，主要从事植物线虫学和大豆抗线虫育种研究，E-mail: jingsheng6673182@163.com。

*通讯作者：李泽宇(1965—　)，男，研究员，主要从事大豆抗病育种和抗性基因研究，E-mail: dqnkylzy@126.com。

population-the SCN race and the Hg (*Heterodera glycines*) type schemes. The objective of this study was to characterize the HG type and SCN race under continuous cropping in daqing and anda. The rescarch results show that: the SCN virulence type of anda is HG type 1. 3. 4. 7 (race 14) and Daqing is HG type 7 (race 3). We reveal race and virulence type of SCN in Daqing and Anda area, provide theoretic foundation for breeding resistant vatieties and controlling of soybean cyst nematode.

Key words: soybean; soybean cyst nematode (*Heterodera glycines*); race; virulence type

我国主要作物上胞囊线虫的种类鉴定及rDNA-ITS分子特征

冯亚星，王东，陈立杰，段玉玺，王媛媛，刘晓宇，陈宇，朱晓峰*

（沈阳农业大学植物保护学院北方线虫研究所，沈阳 110866）

摘　要：胞囊线虫是一类重要的植物寄生线虫，其中很多种类严重威胁着我国粮食生产。2012—2014年，对我国主要作物上的9个胞囊线虫种群进行了形态学和分子鉴定，共鉴定出胞囊线虫4个种，分别为：大豆胞囊线虫（*Heterodera glycines*）、禾谷胞囊线虫（*H. avenae*）、菲利普胞囊线虫（*H. filipjevi*）和旱稻胞囊线虫（*H. elachista*）。首次发现我国安徽省有菲利普胞囊线虫（*H. filipjevi*），江西省有旱稻胞囊线虫（*H. elachista*）的分布。利用PCR-ITS-RFLP方法对胞囊线虫种群进行了鉴别，明确了我国主要作物上胞囊线虫种类的分子特征。

关键词：胞囊线虫；鉴定；rDNA-ITS；RFLP

Identification of cyst-forming nematodes and molecular characteristics based on rDNA-ITS

Yaxing Feng, Dong Wang, Lijie Chen, Yuxi Duan, Yuanyuan Wang, Xiaoyu Liu, Yu Chen, Xiaofeng Zhu*

(College of Plant Protection, Nematology Institute of Northern China, Shenyang Agricultural University, Shenyang 110866, China)

Abstract: Cyst-forming nematode, as the most important one of plant parasitic nematodes, cause heavy yield losses in many crops is a serious threat to food production in China. During 2012-2014, nine populations of cyst nematodes were identified as four species (*Heterodera glycines*, *H. avenae*, *H. filipjevi* and *H. elachista*) with morphological and molecular methods from various locations in major cereal-cultivating areas of China. This is the first report of *H. filipjevi* in Anhui province and *H. elachista*

基金项目：农业部公益性行业科研专项(201503114)；辽宁省教育厅科学技术研究一般项目(L2014273)；辽宁省大学生创业创新训练计划项目(201410157000069)。

作者简介：冯亚星(1993—　)，男，河南新乡人，从事植物线虫分类研究，E-mail：xingxingdiandeng1267@126.com。

王东，男，河北张家口人，博士研究生，主要从事植物线虫学研究，E-mail：wangdong19852008@163.com。

冯亚星和王东为并列第一作者。

* **通讯作者：**朱晓峰，男，主要从事植物寄生线虫研究；E-mail：syxf2000@163.com。

in Jiangxi province. PCR-ITS-RFLP method was applied to distinguish all studied populations, which reveals molecular characteristics of cyst nematode species on major crops in China.

Key words: cyst nematodes; identification; rDNA-ITS; RFLP

南方根结线虫程序性死亡基因 *Mi*PDCD6 的 RNAi 效应分析

祝乐天，陈晨，陈芳妮，张劲蔼，孙思，廖美德，王新荣*
（华南农业大学农学院，广州 510642）

Functional analysis of *Mi*PDCD6 from *Meloidogyne incognita* by RNA interference

Letian Zhu, Chen Chen, Fangni Chen, Jin'ai Zhang, Si Sun, Meide Liao, Xinrong Wang*
(College of Agriculture, South China Agricultural University, Guangzhou 510642, China)

摘　要：根结线虫（*Meloidogyne* spp.）是一类重要的植物病原线虫，给农业生产造成巨大的经济损失。其2龄幼虫侵入寄主根系，食道腺分泌物经口针注入植物细胞，从而刺激寄主植物形成巨型细胞并产生根结。根结线虫食道腺蛋白在线虫与寄主植物互作过程中发挥关键作用。随着分子生物学研究的深入，运用分子生物学技术研究其致病机理以及寻找控制该病害的方法成为可能。本文以南方根结线虫中编码程序性死亡蛋白6（Programmed cell death protein 6，PDCD6）的 *Mi*PDCD6 基因为研究对象，用 RNAi 和半定量 RT-PCR 的方法初步研究了该基因的功能。主要结果为：①半定量 RT-PCR 检测结果表明：被 dsRNA 侵泡后的南方根结线虫 *Mi*PDCD6 基因的转录水平明显降低。②人工接种被 dsRNA 侵泡后南方根结线虫 2 龄幼虫 28 d 后，寄主空心菜（*Ipomoea aquatica* Forsk.）的根结数显著降低。本研究为建立利用 RNAi 技术防控根结线虫的新方法打下基础。

关键词：南方根结线虫；程序性死亡基因 *Mi*PDCD6；RNAi；RT-PCR

基金项目：国家自然科学基金（(30771409 和 31171825)；国家留学基金(2014)。

作者简介：祝乐天，硕士研究生。

* **通讯作者：**王新荣，教授，主要从事植物线虫学研究，E-mail：xinrongw@scau.edu.cn。

3,4,5-三羟基苯甲酸甲酯防治番茄青枯病的作用方式及其对番茄根系次生代谢物质的影响

袁高庆*,陈媛媛,范腕腕,黎起秦,林纬
(广西大学农学院,南宁 530004)

摘　要:3,4,5-三羟基苯甲酸甲酯(methyl gallate,MG)属于酚酸酯类化合物,广泛存在于自然界多种植物中,对多种植物病原菌具有较强的离体抑制活性。本文作者前期研究发现,该化合物可显著影响茄青枯拉尔氏菌生长、形态结构、能量代谢和蛋白质表达等,且稳定性好,在温室及大田条件下均可有效控制番茄青枯病的发生。为进一步明确 MG 有效控制番茄青枯病的机制,为田间合理应用 MG 提供理论支持,该研究配制不同浓度的 MG 药液,在不同时间对水培番茄进行防治试验,通过生物学方法测定 MG 防治番茄青枯病的作用方式;分别在番茄苗期和开花结果期采集自然生长和施用过 MG 的番茄根系,采用气相色谱-质谱联用方法分析该化合物对番茄根系组织次生代谢物质合成的影响。结果表明,3,4,5-三羟基苯甲酸甲酯在番茄植株上表现出较强的内渗作用,这种内渗效果在 MG 浓度较低作用时间较短时即可表现出来。该化合物对番茄青枯病有较好的保护作用,施用 5 d 之内的防治效果超过 87%,且持效期可达 10～15 d,但传导性能弱,对番茄青枯病的治疗效果差。从自然生长和 3,4,5-三羟基苯甲酸甲酯作用过的番茄根系中共检测出 57 种次生代谢物质,3,4,5-三羟基苯甲酸甲酯对部分物质组成和含量有一定影响,受影响比较明显的物质有苯甲醇、水杨酸甲酯、香叶醇、己内酰胺、顺式-11,14-二十碳二烯酸甲酯、亚麻酸、豆甾醇、β-谷甾醇以及木栓酮,其中香叶醇、豆甾醇和 β-谷甾醇这 3 种具有抑菌活性的物质在 MG 处理后表现为含量增加,抑菌活性物质木栓酮为 MG 处理后番茄根系新增的化合物。

关键词:3,4,5-三羟基苯甲酸甲酯;番茄青枯病;作用方式;次生代谢物质

基金项目:广西自然科学基金项目(2013GXNSFAA019056)。

* **通讯作者**:袁高庆,副教授,主要从事植物土传病害及其生物防治,E-mail: ygqtdc@ sina. com。

草莓灰霉病菌拮抗细菌的筛选与初步鉴定

张清霞，谢玮玮，陈夕冉，张璐，陈夕军*

（扬州大学园艺与植物保护学院，扬州 225009）

摘　要：草莓灰霉病是由灰葡萄孢菌（*Botrytis cinerea*）引起的，是草莓生产中最重要的病害之一，也是一种世界性的病害，严重时可使草莓采后损失达50%以上。该病菌可危害花、果实和叶片。贮藏、运输过程中灰霉病菌也是造成采后损失的主要原因。目前，控制草莓采后病害的主要方法仍然是在开花期使用化学杀菌剂，但是往往造成环境污染，且高残留会影响食品质量安全，因此利用拮抗菌进行生物防治具有重要意义。从黑龙江、内蒙古、江苏、浙江等地采集土样60份，分离到细菌809株，通过平板对峙培养初筛，分离到对草莓灰霉菌有较好抑制作用的菌株6个，其中菌株38拮抗能力最强，抑菌带宽0.70 cm，其次是10-3、YS2、24、50和8，抑菌带宽分别为0.63 cm、0.53 cm、0.53 cm、0.51 cm和0.50 cm，抑菌谱实验结果表明10-3、50和8抑菌谱较广。孢子萌发试验结果显示，拮抗细菌10-3、YS2、24、38、50和8的细菌悬浮液对灰霉病菌孢子萌发抑制率均为100%，而经直径0.22 μm微孔滤膜过滤后其无菌发酵液的抑制率分别为90.49%、84.87%、87.37%、71.75%、81.81%、81.26%。其中菌株8还可以产生挥发性化合物，该物质也可完全抑制灰霉病菌分生孢子萌发。灰霉病菌芽管伸长也受到影响，对照处理芽管长度为34.63 μm，而经细菌10-3、38、50、8、YS2和24处理的灰霉菌芽管长度分别为3.29、9.78、6.29、6.49、5.24和4.37 μm。根据细菌16S rDNA序列分析，将菌株10-3初步鉴定为绿针假单胞菌（*Pseudomonas chlororaphis*）。

基金项目：江苏省农业科技资助创新基金 CX（15）1037。

* **通讯作者**：陈夕军，男，副教授，E-mail：xjchen@yzu.edu.cn。

防治果树冠瘿病的农杆菌 K1026 解磷活性研究

魏艳丽[1],Maarten H. Ryder[2],李纪顺[1],李红梅[1],杨合同[1*]
([1]山东省科学院生物研究所,山东省应用微生物重点实验室,济南 250014;
[2] School of Agriculture, Food and Wine, University of Adelaide, SA Australia)

摘　要:从土壤中分离的放射土壤杆菌(*Agrobacterium radiobacter*)K84 菌株是应用较早且可有效防治核果类根癌病的生物制剂,它可以抑制多种有致病性的菌株,而对非致病性的菌株无影响。通过遗传工程技术得到与亲代菌株 K84 有相同染色体背景的衍生菌株 K1026,可避免使病原菌产生抗性,现已在核果类果树生产中广泛应用。本文研究了生防菌株 K1026 对山东泰安地区大樱桃冠瘿病病原菌的抑制作用,并对 K1026 的解磷活性进行初步研究。结果显示:K1026 产生的农杆菌素可抑制绝大多数大樱桃根癌病原菌的生长;胡萝卜切片接种病原菌 0~20 h 后再接种 K1026,可抑制切片瘤状组织的产生,抑制率为 97.8%;在解磷平板上 K1026 可产生直径为 1.32 cm 的溶磷圈,用磷钼蓝比色法在 Pikovskaya 培养基中测定其解磷量,培养 72 h 菌株 K1026 解磷量高达 109.11 μg/mL。说明,生防菌株 K1026 不仅可用于大樱桃根癌病的生物防治,还可用于溶磷微生物肥料的生产。

关键词:冠瘿病;农杆菌;生物防治;解磷作用

基金项目:智惠山东(促生防病微生物农药、肥料的研发)项目。

作者简介:魏艳丽,E-mail:yanli_wei@163.com。

* **通讯作者**

列当生防镰刀菌的筛选及发酵条件的优化

王亚娇，纪莉景，栗秋生，王连生，孔令晓*
（河北省农林科学院植物保护研究所，河北省有害生物综合防治工程技术研究中心，
农业部华北北部作物有害生物综合治理重点实验室，保定 071000）

摘　要：列当是一类恶性根寄生杂草，由于其没有叶绿素，列当生长所需的水分和养分靠用吸盘在寄主根部吸收从而导致寄主因营养不足生长缓慢或枯死，严重影响了农作物的产量和品质。为了开始防治列当的生防菌，从河北张家口采集的弯管列当病株中分离得到了 20 株镰刀菌，通过检测对主栽植物和蔬菜的安全性、对弯管列当种子萌芽抑制作用和对弯管列当的防治作用三种方法筛选获得一株对弯管列当具有明显防治利用的镰刀菌 Br-2。Br-2 菌株对小麦玉米番茄辣椒等植物和蔬菜相对安全且有一定的促生作用，其发酵上清液对弯管列当种子萌芽抑制率为 71.79%，在田间实验中对列当的防效最高可达 75.70%。经过形态学和分子学鉴定 Br-2 菌株为尖孢镰刀菌。为了降低 Br-2 的生产成本，本研究通过单因素试验和正交设计试验对其发酵条件进行了优化，结果表明，Br-2 适宜的发酵培养基为小麦粉 2%、酵母粉 4‰、磷酸氢二钾 2‰、硫酸镁 0.5‰、氯化钾 0.25‰、硫酸亚铁 0.005‰；优化后的培养条件为：初始 pH 7、接种量 1%、培养温度 30℃、摇瓶转速 170 r/min。Br-2 在优化培养基和培养条件下的孢子量提高了 34 倍达到 4.75×10^{8} cfu/mL，大大提高了产孢量，降低了生产成本。

关键词：列当；生物防治；尖孢镰刀菌；发酵优化

* 通讯作者

拮抗木霉 gz-2 菌株在土壤中的空间定殖研究

杜婵娟，付岗*，潘连富，晏卫红

（广西农业科学院微生物研究所，南宁 530007）

摘　要：木霉（*Triochoderma* sp.）是一类广泛存在于土壤中的拮抗微生物，能够拮抗多种植物病原真菌，其生存能力强，适应性广，是一类理想的防治土传病害的生防菌。由于生防菌在作物根围土壤中的定殖情况与其防病效果密切相关。因此，本研究拟以前期获得的 1 株对植物病原真菌具有良好拮抗作用的哈茨木霉（*T. harzianum*）gz-2 菌株作为测试菌株，采用土壤稀释分离法，测定该菌株随时间变化在土壤垂直和水平空间的定殖数量，以期找出该菌株的最佳施用方式和条件，为木霉的田间应用提供科学指导和理论依据。

木霉 gz-2 菌株在土壤垂直空间定殖动态的研究结果表明，土壤深度越深，各深度土层中 gz-2 菌株的数量达到峰值的时间越早。其中，2～4 cm 处该菌株的数量在接种后第 28 天达到峰值，数量为（201.67～228.33）$\times 10^5$ cfu/g；6～14 cm 处该菌株的数量在接种后第 21 天达到峰值，数量为（4.02～27.04）$\times 10^5$ cfu/g；16～26 cm 处该菌株的数量在接种后第 14 天达到峰值，数量为（0.28～3.08）$\times 10^5$ cfu/g；28～40 cm 处该菌株的数量在接种后第 7 天达到峰值，数量为（0.19～0.25）$\times 10^5$ cfu/g。gz-2 菌株在 26 cm 以上土壤深度中的定殖数量呈先上升后下降的趋势，而其在 28 cm 以下土壤深度中的定殖数量则是逐渐减少；56 d 内，其最大定殖深度为 40 cm。

木霉 gz-2 菌株在土壤水平空间定殖动态的研究结果表明，gz-2 菌株的数量在土壤水平方向上总体呈先上升后下降趋势，各定殖半径土壤中 gz-2 菌株的数量均在接种后第 35 天达到峰值，数量为（4.02～27.04）$\times 10^5$ cfu/g，然后便开始逐渐下降。gz-2 菌株在土壤水平方向上呈阶梯式扩散模式，其数量随着定殖半径的增大而减少；63 d 内，其最大定殖直径为 24 cm。

土壤含水量和 pH 对 gz-2 菌株在土壤中定殖数量的影响测试，结果表明，土壤含水量小于 25%时，gz-2 菌株在土壤中的定殖数量随着含水量的增加而逐渐增大；当土壤含水量达到 30%时，该菌株的数量反而减少。而 gz-2 菌株在土壤 pH 为 4.8～8.4 范围内均能生长。当土壤含水量为 25%，pH 为 7.8 时，该菌株在土壤中的数量达到最大值，为 3.38×107 cfu/g。

关键词：木霉；空间定殖；土壤含水量；pH

基金项目：广西农业重点科技计划（No. 201410）；广西农业科学院基本科研业务专项（桂农科 2013YQ03，2015YT79）；南宁市科学研究与技术开发计划项目（201102059）。

作者简介：杜婵娟，硕士，助理研究员，主要从事应用微生物研究，E-mail：duchanjuan@gxaas.net。

* **通讯作者**：付岗，博士，副研究员，主要从事植物病害生物防治研究，E-mail：fug110@gxaas.net。

拮抗葡萄霜霉病生防细菌的筛选及其抑菌效果研究

谢雪迎，王贻莲，扈进冬，杨合同，李纪顺*

（山东省科学院生物研究所，山东省应用微生物重点实验室，济南 250014）

摘　要：本研究从土壤中分离到的 64 株芽孢杆菌为出发菌株，采用微孔板孢子囊萌发试验，筛选得到 8 株芽孢杆菌对葡萄霜霉病孢子囊萌发具有显著抑制作用，其中 BMJBN02 菌株的抑制孢子囊萌发效果最好，萌发率仅 15.49%；而 BCJB01、YB1、YX7 和 YX13 等菌株可将病原菌裂解，其中菌株 BCJB01 可将诱导孢子完全裂解。对抑制葡萄霜霉病孢子囊萌发的菌株进行代谢产物检测，发现菌株 BMJBN02 在纤维素平板、几丁质平板和 β-1,3-葡聚糖平板上产生的水解圈半径最大。鉴于抑菌机制和抑菌效果两个方面考虑，确定 BCJB01 和 BMJBN02 作为重点候选菌株进行菌株种类鉴定。通过对菌株 BCJB01 和 BMJBN02 的形态、生长特性、生理生化反应、16S rDNA 序列测定和系统发育分析，鉴定其分别为蜡质芽孢杆菌（*B. cereus*）和巨大芽孢杆菌（*B. megaterium*）。叶盘法检测发现，经菌株 BCJB01 和 BMJBN02 发酵液处理的叶片，霉层的面积明显减少，发病率分别为 31.1%和 42.3%，病情指数分别为 11.7 和 18.6，抑菌效果显著。经田间试验发现，稀释 50 倍的 BCJB01 悬浮液（菌体含量为 6.6×10^{7} cfu/mL）对葡萄霜霉病具有较好的防治效果，施药后 7 d 降为 16.05，防效可达到 85.92%，控病效果明显优于 0.1%霜霉威。

关键词：葡萄；霜霉病；生防细菌；抑菌

基金项目：公益性行业（农业）科研专项经费（201203035）；山东省重点研发计划（2015GNC113002）。

作者简介：谢雪迎，E-mail：xiexueying111@sina.com。

* **通讯作者**：李纪顺，山东沂水人，高级工程师，主要从事生物农药、生物肥料的研发。

抗重茬菌剂对西瓜土壤微生物群落多样性的影响

高硕，刘正坪，张志勇，赵晓燕*

（北京农学院植物科学技术学院，农业应用新技术北京市重点实验室，北京 102206）

摘　要：土壤微生物是土壤生态系统中很重要的组成部分，对土壤的养分吸收与循环、有机物的降解和能量转化都发挥着重要的作用。抗重茬菌剂是以拮抗细菌为主要成分的微生物菌肥，为明确抗重茬菌剂对西瓜土壤微生物群落多样性的影响，本研究通过使用 Biolog · ECO 平板培养法，对西瓜根部周围的土壤进行了微生物 Biolog 培养，进一步对平均吸光值 AWCD 和多样性指数进行统计分析，并通过分离鉴定的方法进行了土壤微生物多样性及优势种群分析。结果表明：2014 年度在设施西瓜根部施用抗重茬菌剂，一年后，处理田块和对照田块相比，AWCD 值明显增加，说明土壤微生物利用碳源增加，微生物数量增多；处理田块和对照田块相比，物种均匀度 U 增加，物种丰富度 H 基本不变，说明土壤微生物种类没有大的变化，而不同种类之间微生物的量有变化，但这种变化并没有差异显著性，说明土壤微生物群落的多样性没有显著改变。2015 年度我们进一步进行了生长季中土壤微生物多样性的时间动态分析，我们发现，随着时间的变化，平均吸光值 AWCD 在前一个月呈下降趋势，后 3 个月呈上升趋势，并且施用抗重茬菌剂比未施抗重茬菌剂的变化幅度更明显。说明在一个生长季内，西瓜根部土壤中的微生物数量呈现先减少后增多的趋势，可能与西瓜根部分泌物的作用有关；在施用抗重茬菌剂的田块中微生物数量的这种先减少后增多的趋势更加明显，说明抗重茬菌剂对西瓜土壤中微生物的多样性有一定的作用。经分析，施用抗重茬菌剂的田块土壤微生物的物种丰富度和均匀度比对照田块明显下降，说明抗重茬菌剂中的拮抗细菌对土壤中的微生物具有一定的抑制作用。随着时间增长，处理田块和对照田块的平均吸光值 AWCD 均不同程度的增长，最后趋于相等，说明生长后期，抗重茬菌剂中拮抗细菌对土壤微生物的影响变小，此时应该进行补施，加强抗重茬菌剂的影响，更大程度地减少病害的发生。在第 2 个月对土壤微生物进行了分离与鉴定，结果显示在施用抗重茬菌剂的土壤微生物中，81.3%是枯草芽孢杆菌（*Bacillus subtilis*）；18.7%是解淀粉芽孢杆菌（*Bacillus amyloliquefaciens*）；未施重茬剂的土壤微生物中，66.7%是枯草芽孢杆菌（*Bacillus subtilis*）；30.0%是解淀粉芽孢杆菌（*Bacillus amyloliquefaciens*）；3.3%是

基金项目：国家自然科学基金青年基金项目（31201559），现代农业产业体系北京市西甜瓜创新团队建设项目（PXM2013-014207-000103）。

作者简介：高硕，在读硕士，研究方向：植物保护，Tel：010-80794280，E-mail：1435151806@qq.com。

* **通讯作者**：赵晓燕，副教授，研究方向：植物真菌病害综合防治，通信地址：102206 北京市昌平区回龙观镇北农路 7 号，北京农学院植物科技学院，电话：010-80794280，E-mail：zhaoxy777@163.com。

短小芽孢杆菌(*Bacillus pumilus*),说明枯草芽孢杆菌和解淀粉芽孢杆菌是试验地西瓜土壤中的优势种群,其中在施抗重茬菌剂的田块中优势更加明显。

关键词:土壤微生物;抗重茬菌剂;Biolog;生物多样性

重组木霉 L-10 可湿性粉剂贮存稳定性及其防治效果

陈凯，李纪顺，魏艳丽，扈进冬，李红梅，杨合同*

（山东省科学院中日友好生物技术研究中心，济南 250014）

摘　要：重组木霉菌株 L-10 为本室构建，具有潮霉素抗性，其染色体上包含来自巨大芽孢杆菌 Ap25 的 β-1，4-葡聚糖酶基因 *glu*14，亲本菌株为哈茨木霉 LTR-2。本实验利用凹凸棒土载体制备了菌株的可湿性粉剂，并定期检测制剂的活分生孢子含量，对贮存后的制剂进行了温室防治效果试验。结果表明，可湿性粉剂的最佳配方为：分生孢子悬浮液（10 mL）、凹凸棒土（87.5 g）、黄原胶（1 g）、十二烷基苯磺酸钠（1 g）、羧甲基纤维素钠（0.5 g）。制剂初始活分生孢子含量为 13 亿/g，定义保存期内活分生孢子含量不低于 2 亿/g。贮存结果表明：室温条件下，LTR-2 保存期为 9 月，L-10 保存期为 10 月。4℃保存条件下，LTR-2 保存期为 12 月，L-10 保存期为 36 月。温室结果表明，4℃保存 36 月的 L-10 制剂对茄子茎腐病（终极腐霉）有较好的防治作用（$P<0.05$），防治效果达到了 92.9%，较亲本菌株 LTR-2 提高了 53.5%。对番茄猝倒病（致病疫霉）也有较好的防治作用（$P<0.05$），防治效果达到了 91.6%，较亲本菌株 LTR-2 提高了 58.3%。重组木霉菌株 L-10 在 4℃下贮存稳定性较亲本菌株 LTR-2 大大延长，推测原因可能与外源基因的随机插入，改变了某些基因的翻译结果，导致分生孢子细胞壁结构更稳定，具体原因有待进一步研究。另一方面，由于终极腐霉和致病疫霉属卵菌，细胞壁结构主要为纤维素和葡聚糖，而重组木霉菌株 L-10 中包含 β-1，4-葡聚糖酶基因 *glu*14，可有效降解病原真菌的细胞壁，从而导致重组后的 L-10 较亲本菌株 LTR-2 对 2 种卵菌病害的防治效果有显著性提高。

关键词：重组木霉 L-10；可湿性粉剂；贮存稳定性；茄子茎腐病；番茄猝倒病

基金项目：山东省重点研发计划（2015GNC113003）；科技基础性工作专项（2014FY120900）；国家高技术研究发展计划（2011AA10A205）。

* 通讯作者

我国冬麦区小麦赤霉病防治时期研究

徐飞，王俊美，宋玉立*，韩自行，赵凯，刘露露

（河南省农业科学院植物保护研究所，农业部华北南部作物有害生物综合治理重点实验室，郑州 450002）

摘　要：为明确戊唑醇防治小麦赤霉病和减少籽粒中DON（脱氧雪腐镰刀菌烯醇）毒素积累的关键时期。本研究在2014年4月22日至5月4日于河南省焦作市温县试验基地进行小麦花期前后喷施戊唑醇防治小麦赤霉病效果和减少籽粒中DON毒素积累的研究。接种方法：采用小麦扬花盛期喷施小麦赤霉病菌混合孢子悬浮液（5株毒素化学型均为15ADON的禾谷镰刀菌菌株为14LY9-2-4、14AY1-2、14YY1-3、14KF3-8和14ZK1-4），孢子悬浮液浓度为5×10^4个/mL，每平方米喷雾50 mL，接种前先喷水保湿，接种后5 d每天使用手持式电动喷雾机喷清水2次进行保湿以保证发病条件。农药喷施方式：使用手持式小喷壶喷雾，25%戊唑醇可湿性粉剂30 g/亩，用水量为50 L/亩，选择晴天无风条件下，下午4点后喷施。6个处理包括：齐穗期（扬花前2 d）喷药、扬花中期喷药、扬花盛期（扬花后1 d）喷药、扬花后4 d喷药、扬花后10 d喷药和不喷施农药对照，另外扬花盛期喷药时间在接种之前。结果表明：扬花中期和扬花盛期进行农药防治效果最好，小麦赤霉病发病率、病小穗率、千粒重、病粒率和籽粒中DON毒素积累分别为52%～64%、14%～21%、44～45 g、7.6%～8.1%、7.6～8.1 mg/kg，不施农药对照的为85%、37%、37 g、18.7%和19.1 mg/kg，两者有显著差异；扬花后4 d和扬花后10 d进行农药防治，在各个指标上防治效果和不喷施农药对照没有显著差异；齐穗期进行农药防治，虽然在赤霉病发病率和病小穗率上与对照没有显著差异，但是在千粒重、病粒率和籽粒中DON毒素积累指标上效果较好。进一步试验仍在进行中。

关键词：小麦；小麦赤霉病；防治时期；脱氧雪腐镰刀菌烯醇

基金项目：公益性行业科研专项（201303016）；“十二五”国家科技支撑计划（2011BAD16B07）；河南省小麦产业技术体系（S2010-01-05）。

作者简介：徐飞，男，湖北麻城人，博士，助理研究员，主要从事小麦病害研究，E-mail：xufei198409@163.com。

* **通讯作者**：宋玉立，研究员，主要从事小麦病害研究，E-mail：songyuli2000@126.com。

内生恶臭假单胞菌 JD204 对小麦条锈病的防治效果及提高产量的影响

庞发虎[1,2]，王坦[2]，黄思良[2*]，余自伟[2]

([1]广西大学农学院，南宁 530005；[2]南阳师范学院生命科学与技术学院，南阳 473061)

摘　要：小麦条锈病是由 *Puccinia striiformis* West. f. sp. *tritici* Eriks 引起的真菌性病害，由于其发生区域广，危害损失重等特点成为影响小麦稳产高产的主要病害。作者于 2014 年 10 月用分离自小麦根部的内生恶臭假单胞菌 JD204 悬浮液(10^7 cfu/mL)浸泡 14 个不同品种小麦种子后分别播种于南阳国家农业园区试验地(经度 112°27′49″；纬度 32°57′6″)中，每小区面积为 4 m^2(播种量 70 g)，每处理 3 个重复，以清水浸泡麦种为对照。于 2015 年 5 月上旬，对小麦自然感染条锈病的发病情况、产量进行调查和统计。结果表明，14 个供试小麦品种中，用 JD204 内生菌处理后平均发病率和病情指数分别为 15.27%和 4.90，分别为对照处理的 68.9%和 64%，分别有 9 个和 11 个小麦品种发病率和病情指数与对照相比显著性的降低；对 14 个不同品种条锈病的防治效果为－98.9%～63.7%。在 7 个高抗品种中，3 个品种(开麦 20、新原 958、濮麦 9 号)为正的防治效果，为 26.1%～53.3%(平均 38.9)，其余 4 个品种(矮优 66、金丰 3 号、中育 12、众连 1 号)为负的防治效果，为－39.4%～－98.9%(平均为－59.35%)。在 7 个高感、中感、中抗品种(高优 503、豫麦 69、豫农 416、郑麦 9023、偃高 006、矮抗吨产王和豫麦 130)上为正防治效果，达 35.0%～63.7%(平均 49.3%)。4 个品种(高优 503、新原 958、豫农 416 和豫麦 130)的防治效果为 53.3%～63.7%，显著性地高于其他 6 个品种(开麦 20、濮麦 9 号、豫麦 69、郑麦 9023、偃高 006 和矮抗吨产王)。有 12 个品种(豫农 416、偃高 006、矮抗吨产王、中育 12、高优 503、濮麦 9 号、郑麦 9023、开麦 20、众连 1 号、矮优 66、新原 958 和豫麦 130)的产量比对照分别提高了 2.0%～20.2%(平均 10.5%)，2 个品种(豫麦 69 和金丰 3 号)产量与对照相比，分别减产了 10.1%和 2.0%。表明内生假单胞菌 JD204 对小麦条锈病的防治和提高产量均具有一定的影响。

关键词：小麦；内生细菌；条锈病；产量

基金项目：河南省高校科技创新团队支持计划项目(2010JRTSTHN012)。

作者简介：庞发虎(1975—　)，男，博士研究生，研究方向为植物病害生物防治，E-mail：pangfahu@163.com。

* **通讯作者**：黄思良，E-mail：silianghuang@126.com。

木霉拮抗灰霉菌与pH的相关性分析

吴晓青[1],赵忠娟[1],赵晓燕[1],陈凯[1],李纪顺[1],杨合同[1,2]*
([1]山东省科学院中日友好生物技术研究中心,山东省应用微生物重点实验室,济南 250014;
[2]山东理工大学,淄博 250049)

摘　要:灰霉菌(*Botrytis cinerea*)可引起果蔬花卉的灰霉病病害,一些木霉(*Trichoderma* spp.)菌株可拮抗灰霉菌并具有防治灰霉病的作用。研究表明,灰霉菌侵染植物初期向环境中分泌草酸(Oxalic Acid)毒素,分泌的草酸通过降低 pH 等途径引起和增强病原菌的致病性。我们前期的研究表明,哈茨木霉(*T. harzianum*)LTR-2 具有防治灰霉病和消除草酸的能力,其与灰霉菌的对峙培养可提高环境 pH。本文进一步分析了58 株木霉菌株拮抗灰霉菌的能力与环境 pH 变化之间的关系,这些木霉菌株由采集自山东、四川、江苏等 18 个省(直辖市)的土样(少数为水样)中筛选获得。在 25℃暗培养条件下,将木霉与灰霉菌进行 PDA 平板上的对峙培养,设 3 个重复。共培养 4 d 后,计算木霉对灰霉菌的抑制率[抑制率=$(D-d)/D\times100\%$,其中 D 为对照灰霉菌生长直径,d 为对峙病原菌生长直径且木霉覆盖生长的部分不计入],同时取距灰霉菌接种点 1 cm 处的 10 处培养基菌块,用 pH 精密试纸(生工®)分别检测 pH,计算平均值。使用 OriginPro8.5 软件对数据进行 Pearson 相关性分析。培养基的初始 pH 为 6.5,同样条件放置 4 d 后 pH 不变。单独培养灰霉菌 4 d 后,接种点附近的 pH 下降至 3.3。灰霉菌与 58 株木霉进行对峙培养 4 d 后,灰霉菌接种点附近的 pH 最高为 7.5、最低为 3.0。经测定,木霉对灰霉菌的抑制率最高为 90.56%、最低为 1.12%。Pearson 相关性表明,木霉对灰霉菌的抑制率越高,pH 回升的幅度越大。Pearson 相关系数 r 值为 0.856 63,可线性拟合,拟合系数 Adj. R-Square 为 0.728 97。两者呈显著相关,并且高度线性相关。上述结果表明在木霉对灰霉菌的抑制率越高,伴随着环境 pH 的升高。这种现象推测与木霉消除草酸的作用有关,暗示消除草酸可能是木霉防治灰霉病的关键机制之一。

关键词:木霉;灰霉菌;pH;Pearson 相关性

基金项目:国家科技基础性工作专项(2014FY120900);山东省中青年科学家科研奖励基金(BS2015SW029);山东省科学院青年基金项目(2014QN019)。

作者简介:吴晓青,E-mail:xq_wu2008@163.com。
*通讯作者

耐盐木霉菌株的分离鉴定及其抗菌促生作用

赵晓燕[1],陈凯[1],吴晓青[1],赵忠娟[1],李纪顺[1],杨合同[1,2*]

([1]山东省科学院中日友好生物技术研究中心,山东省应用微生物重点实验室,
济南 250014;[2]山东理工大学,淄博 250049)

摘要:2014 年本实验室从山东东营、潍坊、海阳、威海、黄海等滨海盐地采集了 120 个盐渍土样品,采用稀释平板法从这些盐渍土样中分离出 83 个木霉菌株。耐盐菌株的筛选方法为 PDA 中加入 2.3%的 NaCl, 25℃培养 72 h,测量菌落半径,以 PDA 培养基为对照,计算耐盐率(%)[耐盐率=100×处理菌落半径/对照菌落半径×100%]。在含 2.3%NaCl 的 PDA 平板上筛选出耐盐率高于 60.00%的木霉菌株 10 个,这 10 个菌株的生长量相当于无 NaClPDA 培养基的 60.00%~81.33%。10 个耐盐菌株采用形态学、ITS 以及 TEF 序列分析完成鉴定,其中 5 株为哈茨木霉(*Trichoderma harzianum*),2 株为长枝木霉(*T. longibrachiatum*),2 株为棘孢木霉(*T. asperellum*),1 株为盖姆斯木霉(*T. gamsii*)。耐盐率最高(81.33%)的是盖姆斯木霉 TW20029,采自山东威海滨海滩涂盐渍地中,这也是我省首次发现 *T. gamsii*,本文进一步研究了 TW20029 的抗病促生作用。采用 PDA 平板对峙培养的方法,计算 TW20029 对 4 种病原菌的抑制作用[抑制率=(*D*-*d*)/*D*×100%,其中 *D* 为对照灰霉菌生长直径,*d* 为对峙病原菌生长直径且木霉覆盖生长的部分不计入],结果表明 TW20029 对水稻立枯病病菌(*Rizoctonia solani*)室内 72 h 抑制率为 70.13%,对终极腐霉(*Pythium ultimum*)室内 96 h 抑制率为 78.72%,对黄瓜枯萎病(*Fusarium oxysporum*)和黄瓜灰霉病(*Botrytis cinerea*)室内 96h 抑制率均为 100%,说明 TW20029 对这 4 种病原菌均具有高效的抑制作用。通过室内种子萌发实验测定了 TW20029 孢子悬浮液(10^8 cfu/mL)对黄瓜种子的发芽影响,实验结果表明 TW20029 孢子悬浮液能够显著提高黄瓜种子的发芽率(85.00%)、发芽指数(11.24)和活力指数(9.02),而水对照的发芽率仅为 60.33%,发芽指数为 9.07,活力指数为 6.00,可以看出 TW20029 对黄瓜种子的发芽具有较强的促进作用。

关键词:木霉;耐盐;抑菌;发芽

基金支持:国家科技基础性工作专项(2014FY120900);山东省中青年科学家科研奖励基金(BS2015SW029);山东省自然科学基金项目(ZR2015PC009);山东省科学院青年基金项目(2014QN019 和 2014QN017)。

作者简介:赵晓燕,E-mail:zhaoxy @sdas. org。

* **通讯作者**

1,3-二氯丙烯熏蒸土壤对病虫草害的防效评价

Evaluation of 1,3-Dichloropropene fumigation for the control of soil-borne pests

刘秀梅,程星凯,乔康*,王开运

(山东农业大学,泰安 271018)

摘　要:我国是重要的设施蔬菜生产国,因长期连作种植导致以根结线虫为代表的土传病虫害发生严重,制约了设施蔬菜的发展。实践证明,使用甲基溴(methyl bromide)熏蒸处理土壤是防治土传病虫害最有效的方法。然而,甲基溴作为一种臭氧层消耗物质,将于2015年1月1日在我国禁用。因此,寻找甲基溴替代品势在必行。1,3-二氯丙烯(1,3-dichloropropene)是一种很有潜力的甲基溴替代物。本课题组通过室内毒力试验和大田验证试验,研究了1,3-二氯丙烯熏蒸土壤防治南方根结线虫、杂草种子和土传病害病原菌的效果,分析其在我国保护地蔬菜上应用的可行性。

采用直接触杀法测定了1,3-二氯丙烯对南方根结线虫的毒力。结果表明,1,3-二氯丙烯对南方根结线虫的LC_{50}和LC_{90}分别为1.20 mg/L和3.74 mg/L。采用美国农业部杂草种子处理方法研究了1,3-二氯丙烯对多种杂草种子的剂量-响应关系。结果表明,杂草种子对1,3-二氯丙烯敏感性由大到小顺序为:马唐>牛筋>稗草>反枝苋,其LC_{50}在14.23~73.59 mg/kg。采用十字交叉法测定了1,3-二氯丙烯对辣椒疫霉病菌、草莓枯萎病菌、棉花立枯病菌、烟草黑胫病菌和番茄灰霉病菌的毒力。结果表明,1,3-二氯丙烯对辣椒疫霉病菌和草莓枯萎病菌的LC_{50}分别为0.24 g/m^{2}和1.55 g/m^2,1,3-二氯丙烯熏蒸对辣椒疫霉病菌最为敏感,其他种类病原菌则表现出中等程度的敏感性。

分别在温室大棚番茄、黄瓜、大姜作物上进行大田试验来验证1,3-二氯丙烯(90 L/hm^2、120 L/hm^{2}和180 L/hm^2)对南方根结线虫、杂草和土传病害病原菌的防治效果。结果表明,与对照组相比,1,3-二氯丙烯施用后能够明显促进作物生长,增强植株活力,有效抑制根结线虫侵染和种群数量,降低根结指数,减少土传病害发生率,增加作物产量。并且中高剂量的1,3-二氯丙烯熏蒸处理在除杂草防治以外的各种防治指标上达到甚至超过甲基溴处理的防治水平,在作物产量上与甲基溴处理之间无显著性差异。

上述研究成果表明1,3-二氯丙烯熏蒸土壤防治蔬菜根结线虫效果良好,并可控制一些土传病害发生,是一种很有潜力的甲基溴替代物。但是,1,3-二氯丙烯对杂草的防治效果一般。因此,建议将1,3-二氯丙烯与其他化学替代品或非化学替代技术结合使用,以达到综合防治的目的。同时,本研究成果能够为1,3-二氯丙烯在设施蔬菜上的合理使用、克服土壤连作障碍和保障设施蔬菜可持续生产提供理论依据。

关键词:土壤熏蒸;1,3-二氯丙烯;根结线虫;土传病害;杂草

---

* **通讯作者**:乔康,E-mail:qiaokang11-11@163.com;qiaokang@sdau.edu.cn。

CRISPR/Cas9 系统敲除水稻基因的研究

杨芳，孙毅，黄寿光，陈旭君*

（中国农业大学农学与生物技术学院，北京 100193）

摘　要：CRISPR/Cas9 系统是近年来开发的一种高效的核酸定点编辑技术，在很多物种中进行着实验。本研究利用 CRISPR/Cas9 系统，高效地对水稻基因进行敲除，为研究基因功能提供了良好的材料。利用网站 CRISPR-Plant 设计靶序列，构建到 Cas9 载体，转化农杆菌后，侵染水稻愈伤。随后筛选抗性愈伤，培养出转基因植株，并检测转基因水稻苗的突变类型及效率。分析检测结果发现，转基因 T0 代植株突变率高达 98%，主要以插入突变为主，占 76.1%，其中 70% 为单碱基插入；缺失突变占 23.9%，缺失碱基长度在 1～17 碱基对之间。同时随机挑选了几个基因进行脱靶检测，没有发现脱靶现象。利用两个农杆菌共侵染愈伤，获得了预期的双敲除突变体，双敲除效率高达 75%。此外，在同一个载体中设计了针对同一个基因的两个靶位点，在对其抗性愈伤的检测中发现在两个靶位点之间发生了大片段的缺失，也存在缺失位置不完全在两个靶位点之间。由此可见利用 CRISPR/Cas9 系统能够高效的获得水稻突变体。

关键词：CRISPR/Cas9；基因编辑；水稻

作者简介：杨芳，硕士研究生，E-mail：yfwe928@163.com；孙毅，硕士研究生，E-mail：wdsunyi@163.com。

杨芳和孙毅为并列第一作者。

* **通讯作者**：陈旭君，副教授，E-mail：chenxj@cau.edu.cn。

病原真菌纤维素酶保守的结构域涉及激发植物的防卫反应

马亚男，陈进银，李多川*
（山东农业大学，泰安 271018）

摘　要：目前，植物病原真菌产生的纤维素酶、木聚糖酶和果胶酶在植物与真菌互作机制方面引起人们的兴趣和关注。我们的研究发现：在植物与真菌互作中，玉米纹枯病菌（*Rhizoctonia solani*）的一种45家族内切纤维素酶EG1具有激发子（elicitor）功能，诱导植物的防卫反应（植物过敏性细胞死亡、植物防卫反应基因表达、活性氧产生、培养基碱化、钙离子积累、乙烯合成等），并且证明它们的激发活性与酶活性没有关系（Ma et al.，2015）。为了进一步理解真菌纤维素酶EG1与植物互作的分子机制，我们通过定点突变的技术，证明真菌纤维素酶EG1保守的结构域GCNWRFDWF涉及它的激发活性。当GCNWRFDWF突变后，尽管EG1有纤维素酶的活性，但它的激发活性丧失。研究结果为进一步分离EG1在植物中的受体蛋白奠定了基础。

关键词：真菌纤维素酶；激发子；激发活性位点

参考文献

Ma Y N, Han C, Chen J Y, *et al*. Fungal cellulase is an elicitor but its enzymatic activity is not required for its elicitor activity. Molecular Plant Pathology, 2015, 15:14-26.

基金项目：国家自然科学基金项目（31071732）；“863”计划（2012AA10180402）。

* 通讯作者

灰葡萄孢弱致病力菌株 HBtom-372 中相关真菌病毒的研究

郝芳敏，张静，杨龙，李国庆，吴明德*

（华中农业大学植物科学技术学院，武汉 430070）

摘　要：由灰葡萄孢（*Botrytis cinerea*）引起的灰霉病是一种世界性分布广泛的真菌病害，可造成多种果蔬的经济损失。对于灰霉病的防控，由于缺乏有效的抗性品种及病菌易对杀菌剂产生抗性，所以生物防治灰霉病将成为一种可能的替代防治措施。前人的研究表明真菌病毒对栗疫病有一定生防潜力，所以从灰葡萄孢群体中筛选获得可导致弱毒相关的特性的灰葡萄孢菌株，将有可能用于未来灰霉病的生物防治中。本研究从湖北荆门分离获得一株弱致病力灰葡萄孢菌株 HBtom-372。该菌株对离体的烟草叶片致病力致病力弱，在 PDA 培养基上生长速率十分缓慢，菌落形态异常，长期培养没有菌核及分生孢子的产生。通过对菌株 HBtom-372 菌丝中 dsRNA 的提取检测，我们发现菌株 HBtom-372 菌丝中含有 7 条 dsRNA 条带，依大小分别命名为 A1、A2、B、C、D、E 和 F。通过对这 7 条 dsRNA 进行了序列的克隆和分析，结果表明 HBstr-372 可能被多种真菌病毒复合侵染，其中片段 A1 可能是一种 endorvirus，大小为 11 557 bp，BLAST 分析表明其与 *Sclerotinia sclerotiorium* endornavirus 1 的同源关系最高。片段 A2 可能为 hypovirus，大小为 10 252 bp，BLAST 分析表明其与 *Sclerotinia sclerotiorium* hypovirus 1 的同源性最高，而 C 和 F 是 A2 的卫星病毒，片段 B 与 fusarivirus 亲缘关系较近，大小为 8 445 bp，BLAST 分析表明其与 *Sclerotinia sclerotiorium* fusarivirus 1 的同源性最高，而片段 D 和 E 是 B 的缺陷性病毒。将弱毒菌株 HBtom-372 与强毒菌株 HBtom-459 进行对峙培养发现，其中的真菌病毒可以通过菌丝融合传给菌株 HBtom-459。通过对传染后的衍生菌株进行研究表明，dsRNA 片段 A2 和 B 成功传染到菌株 HBtom-459 中，且感染这两条片段的衍生菌株表现出致病力衰退及生长速度减慢的特征。因此，我们认为 A2 和 B 这两条片段可能与灰葡萄孢菌株 HBtom-372 的弱毒特性密切相关。

关键词：灰葡萄孢；灰霉病；真菌病毒；dsRNA

基金项目：行业专项“保护地果蔬灰霉病绿色防控技术研究与示范”（201303025）。

* **通讯作者**：吴明德，博士，E-mail：mingde@mail.hzau.edu.cn。

植物内生菌对柑橘溃疡病的抑菌活性及生物学性状分析

金玲莉*，涂娟，王彦波，曾明，杜玉标
（江西省农业科学院园艺研究所　330200）

摘　要：从南丰蜜橘叶片和洋葱的健康植物组织提取液中筛选出3株对柑橘溃疡病（*Xanthomonas campestris* pv. *citri*（Hasse）Dye）有拮抗作用的内生菌株Bb1、Bb2和YC1，通过拮抗菌株的抑菌活性试验得出3株菌株对柑橘溃疡病均有较好的抑制效果，并对其生物学性状进行了研究分析。

关键词：拮抗；内生细菌；抑菌活性；形态特征；生物学性状

Biological characteristics of plant endophyte on citrus bacterial canker disease and its antimicrobial activity

Abstract: From the nanfeng Orange leaves and onions healthy plant tissue extracts screened in 3 strains of citrus Bacterial canker disease (*Xanthomonas campestris* pv. *citri*(Hasse)Dye) there is antagonism of endophytic strains Bb1, andBb2 and YC1, obtained through the antagonistic strains antimicrobial activity test 3 strains of citrus canker has very good inhibitory effect, and its biological characteristics were studied.

Key words: antagonism; endophytic bacteria antimicrobial activity; features; biological characters

植物内生菌是指其生活史的某一阶段或全部阶段能在健康植物组织内栖居而对植物不造成实质性危害并与植物建立了和谐联合（compatible association）关系的微生物，主要包括真菌、细菌和放线菌，能有效地抑制病原菌的侵染或提高宿主植物的抗病性[1]。

内生细菌是植物各种组织和器官的细胞间隙或细胞内的天然宿居者，在其部分或整个生活周期寄居在植物组织，特别是营养繁殖的组织中生存、繁衍、传播[2]。由于植物内生细菌与宿主协同进化，在演化过程中形成了明显的互惠共生关系。许多研究表明，植物内生细菌对宿主具有抗病促生作用，显示了其在生物农药开发和促进植物生长等方面有着良好的应用前景[3]。

项目来源：江西省科技厅支撑计划项目，项目编号：20132BBF60029，投稿日期：2015-01-20。

* **作者简介**：金玲莉（1976—　），女，副研究员，主要从事病虫害防治研究。E-mail：jinlinglilove@163.com，联系电话：13870664239。

1　材料与方法

本研究采用研磨分离法从南丰蜜橘健康组织叶片和洋葱提取液中筛选出3株对柑橘溃疡病有拮抗作用的内生菌株，用抑菌圈法测定其抑菌效果，并对其生物学性状进行分析研究。

南丰蜜橘叶片取至江西省农业科学院园艺研究所的柑橘资源圃内，洋葱来自超市购买的新鲜无病虫害洋葱。在园艺研究所3 000 m^2 的南丰蜜橘资源圃内，采用5点取样法，选取生长性状良好，无病害症状的健康植株20株，在每株树中采取健康叶片5片，采集后立即放入无菌样品袋中，带回实验室进行内生细菌的分离。将初步筛选出的拮抗菌接柑橘溃疡病菌，于生化培养箱内培养，用抑菌圈法测定其抑菌效果，利用公式计算抑菌率。

NA培养基：牛肉膏3 g，蛋白胨5 g，葡萄糖2.5 g，琼脂17 g，水1 000 mL，pH 7.2。

2　结果与分析

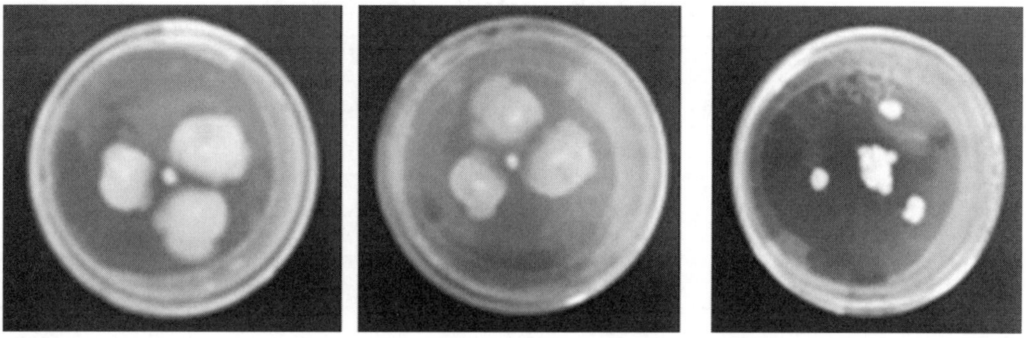

Bb1　　Bb2　　YC1

图1　拮抗菌株抑菌效果图

2.1　拮抗内生菌抑菌活性

表1　分离的内生细菌性状描述

菌株 Bacterial strain	来源 Ovigin	颜色 Colour	荧光 Fluorescence	抑菌圈直径(mm) Growth inhibition zone
Bb1	南丰蜜橘	黄色	无	14.0
Bb2		浅黄色	无	9.0
S1	蒜	浅黄绿色	无	0
S2		浅黄绿色	无	0
S3		白色	无	0
S4		白色	无	0
S5		乳白色	无	0
S6		黄色	无	0

续表 1

菌株 Bacterial strain	来源 Origin	颜色 Colour	荧光 Fluorescence	抑菌圈直径(mm) Growth inhibition zone
C1	葱	白色	无	1.4
C2		黄绿色	无	0.9
YC1	洋葱	乳白色	无	25
YC2		黄色	无	0

从图 1 和表 1 可以看出：从南丰蜜橘叶片中提取的内生菌 Bb1、Bb2 和从洋葱提取液中筛选的内生菌 YC1 对柑橘溃疡病有明显的抑制作用。而从葱中提取的内生菌 C1、C2 对柑橘溃疡病有较弱的抑制作用，其他内生菌均没有抑制作用。从抑菌圈直径看，YC1 最大，其次为 Bb1，Bb2 比 Bb1 相对较弱，从葱提取液中提取的内生菌抑菌圈直径很小，其他都为零。

表 2　3 株拮抗菌悬浮液对柑橘溃疡病的抑制效果

菌株种类	抑菌圈面积(cm^2)				
	重复 1	重复 2	重复 3	平均面积	抑菌率
Bb1	5.12	4.38	4.97	4.82	16.56
Bb2	7.39	6.67	5.84	6.63	15.43
YC1	7.30	8.55	5.31	7.05	17.82
CK	0.00	0.00	0.00	0.00	0.00

从表 2 可以看出：3 株拮抗菌中，YC1 的抑菌率最大，拮抗作用最强，其次是 Bb1，最后是 Bb2。

2.2　拮抗菌株生物学性状分析

2.2.1　个体形态性状

Bb1 菌体：杆状，营养体大小为 3.1 μm×1.0 μm，芽孢大小为 1.9 μm×0.8 μm，周生鞭毛，芽孢位于菌体中间，革兰氏阳性。

Bb2 菌体：杆状，大小为 3.1 μm×0.7 μm，芽孢中生，芽孢大小为 1.6 μm×0.6 μm，周生鞭毛，革兰氏阳性。

YC1 菌体：杆状，大小为 2. 83 μm×0.71 μm，极生鞭毛 1～2 根；革兰氏染色阳性，无芽孢，无荚膜。

2.2.2　菌落形态性状

Bb1 菌落：菌落培养时中央隆起，直径为 0.3 cm，表面光滑，整齐，不透明，圆形，颜色为黄色。

Bb2 菌落：菌落培养时为隆起状，直径为 0.5 cm，表面光滑，整齐，不透明，近圆形，颜色为浅黄色。

YC1 菌落：菌落培养时为不规则堆状，边缘呈波浪形，表面光滑，不整齐，不透明，颜色为乳白色。

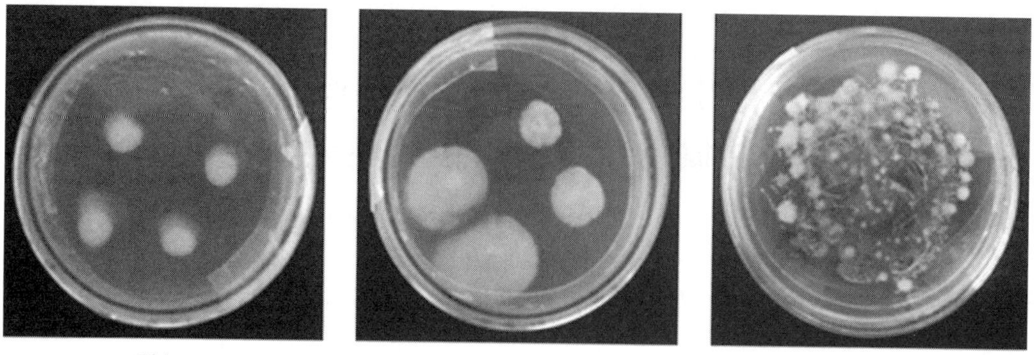

Bb1　　Bb2　　YC1

图 2　菌落形态

参考文献

[1]　史应武，娄恺，等. 植物内生菌在生物防治中的应用[J]. 微生物学杂志，2009，29(6)：61-64.

[2]　张颖，王刚，郭建伟，等. 利用小麦内生细菌防治小麦全蚀病的初步研究. 植物病理学报，2007，37(1)：105-108

[3]　高增贵，庄敬华，陈捷，等. 玉米根系内生细菌种群及动态分析. 应用生态学报，2004，15(8)：1344-1348.

中国小麦花叶病毒 CP 和 CRP 蛋白的原核表达、抗血清制备及 RNA2 侵染性克隆构建

孔凡惠[#1]，脱建波[#1]，魏娇[1]，唐伟[1]，李向东[1*]，迟胜起[2]，田延平[1]，于金凤[1]
([1] 山东农业大学植物保护学院植物病理学系，泰安 271018；
[2]青岛农业大学农学与植保学院，青岛 266109)

摘　要：中国小麦花叶病毒(*Chinese wheat mosaic virus*，CWMV)引起的小麦土传花叶病在山东烟台、威海等地危害严重。本研究克隆了 CWMV 烟台分离物的外壳蛋白(Coat protein，CP)及富含半胱氨酸蛋白(Cysteine-rich protein，CRP)基因，并将其连接到原核表达载体 pEHISTEV，转化大肠杆菌 Rosetta。经 IPTG 诱导，表达出分子量均为 19 kDa 的 CP 和 CRP。将二者从凝胶中切下，乳化后免疫新西兰大耳兔 4 次，获得了两种蛋白的多克隆抗体。ELISA 检测表明，CWMV CP 和 CRP 抗血清的效价分别为 1∶4 096 和为 1∶2 048。Western blot 分析证明该抗血清只与感染 CWMV 的小麦有特异性反应，而与健康小麦或感染小麦黄花叶病毒的小麦无反应。利用含 T3 启动子的引物通过 RT-PCR 扩增 CWMV RNA2 全长片段，经 T/A 克隆连接到 pMD18-T，获得质粒 pMD18-T-CWMV-RNA2。该质粒经 Xba I 线性化后，利用 T3 RNA 聚合酶进行体外转录，转录产物摩擦接种本氏烟，15℃培养 3 d 后，利用 Western blot 可从接种叶片中检测到瞬时表达的 CWMV CP 蛋白。

关键词：中国小麦花叶病毒；衣壳蛋白；富含半胱氨酸蛋白；抗血清制备；侵染性克隆

小麦黄花叶病毒衣壳蛋白的原核表达及抗血清制备

唐伟[1#]，程德杰[1#]，魏娇[1]，孔凡惠[1]，李向东[1,2*]，于金凤[1,2]

（[1] 山东农业大学植物保护学院植病系，泰安 271018；

[2]山东省小麦玉米周年高产高效协同创新中心，泰安 271018）

摘　要：通过 RT-PCR 的方法扩增获得小麦黄花叶病毒（*Wheat yellow mosaic virus*，WYMV）的衣壳蛋白（CP）基因，并将其连接原核表达载体 pEHISTEV。将重组质粒转化到大肠杆菌 Rosetta，通过 IPTG 诱导后，可以表达 38 kDa 的融合蛋白。通过切胶回收的方法收集目的蛋白，免疫新西兰长耳兔，制备 WYMV CP 的多克隆抗体。间接 ELISA 测定该抗血清效价为 1∶2 048，Western blotting 分析表明该血清可以与病株中 CP 发生特异性反应，而与健康植株汁液无反应。

关键词：小麦黄花叶病毒；外壳蛋白；原核表达；抗血清；检测

基金项目：农业公益性行业科技项目（201303021）；山东省现代农业产业技术体系。

Characterization of *Chinese wheat mosaic virus* isolates from Shandong province

Wu Bin, Xin Xiangqi, Jiang Shanshan, Zhang Mei, Wang Shengji, Zhao Jiuhua*

(Institute of Plant Protection, Shangdong Academy of Agricultural Sciences, Jinan, 250100)

Abstract: *Chinese wheat mosaic virus* (CWMV) belongs to the genus *Furovirus*, and is naturally transmitted by *Polymyxa graminis* L., which is widely distributed in soil. CWMV often co-infects with Wheat yellow mosaic virus, negatively effecting crop production on an annual basis. To gain insights into the evolutionary mechanisms of CWMV, the complete RNA genome sequences of 7 CWMV isolates from Shandong Province were determined. Phylogenetic analyses revealed that CWMV isolates can be divided into two evolutionary divergent groups based on CP and MP, respectively. Recombination was found in one of the seven isolates and the recombination site was sometimes located between two coding gene sequences, indicate that different coding genes of the isolate may evolve separately, consistent with modularity in evolution. The results of the selection pressure and population diversity analysis suggest the negative selection of all the CWMV coding genes and the long-term stability of the population.

Key words: *Chinese wheat mosaic virus* (CWMV); complete sequence; phylogenetic analysis; recombination

---

Foundation project: Special Fund for Agro-scientific Research in the Public Interest (201303021).

State Key Laboratory Breeding Base for Zhejiang Sustainable Pest and Disease Control (No. 2010DS700124-KF404).

First author: Wu Bin(1986-), male, Shandong Liaocheng, Doctor, Research Associate, the main direction is plant virology, Email: wubin228@126.com.

* Corresponding author: Zhao Jiuhua, professor, the main direction is plant virology, Email: zhaojiuhua2009@163.com.

新型药剂对花生防病增产试验

曹欣然，田明英*

（烟台市植保站，烟台 264001 ）

摘　要：[目的]通过试验，验证新型药剂对花生主要病害的防治效果和增产效果，为进一步大面积推广提供理论依据。[方法]田间试验地块设在烟台市牟平区，花生品种为青花 7 号。采用药剂拌种，共 6 个处理，分别为处理一（25%噻虫.咯.霜灵悬浮种衣剂(FS)350 mL/百千克种子拌种）、处理二（48%溴虫.噻虫嗪 FS 400 mL/百千克种子+25%嘧菌酯悬浮剂(SC)50 mL/亩拌种+25%咯菌腈 FS 150 mL/百千克种子）、处理三（48%溴虫、噻虫嗪 FS 500 mL/百千克种子+25%嘧菌酯 SC 50 mL/亩拌种+25%咯菌腈 FS 150 mL/百千克种子）、处理四（48%溴虫·噻虫嗪 FS 600 mL/百千克种子+25%嘧菌酯 SC 50 mL/亩拌种+25%咯菌腈 FS 150 mL/百千克种子）、处理五（60%吡虫啉 SC 200 mL/百千克种子+40%萎莠·福美双 FS 170 mL/百千克拌种）、处理六（31%氨基·噻虫嗪·氟虫腈 FS 40 mL/亩）与空白对照，分别在幼苗期、开花下针期和荚果期进行叶面药剂喷雾。播种后 15 d 调查出苗率；近收获期调查花生叶斑病发生情况，计算病情指数和防治效果；花生收获时测产。[结果]出苗率：出苗时间上拌种处理比常规对照晚出苗 1 d。出苗率上 6 个处理与常规对照在 89.47%～93.69%之间，没有显著性差异。花生叶斑病：药剂处理区病情指数分别在 2.00～3.11之间，防治效果在 93.70%～95.94%之间，各个处理区之间没有显著性差异，均显著优于常规对照区。产量：处理一亩产 378.78 kg，比对照区增产 22.48%；处理四亩产 363.00 kg，增产 18.31%；处理六亩产 360.80kg，增产 17.60%；处理五亩产 354.13 kg，增产 15.42%；处理三亩产 337.70 kg，增产 10.07%；常规对照亩产 306.81 kg。通过方差分析，处理四和处理六之间没有显著性差异，其余各处理间差异性显著。[结论]①田间调查表明，拌种花生处理区较未拌种花生处理区虽然晚出苗一天，但出苗率与常规对照没有显著性差异，从而保证花生高产所需的亩株数。②收获期调查表明，处理区较常规对照区绿叶多，病害发生程度低，基本无落叶，而常规对照区叶斑病发生程度高，病叶率达到 100%，在收获前 14 d 开始大量落叶。说明处理区对花生叶斑病有显著的防治效果。因为能够有效控制叶斑病发生程度，保证花生灌浆所需时间，处理区产较常规对照区量提高 5.16%～22.48%，增产效果非常显著。根据结果，建议在大田推广中轮换应用处理一、处理四、处理六和处理五，避免产生抗药性。

关键词：花生；病害；产量；新农药

作者简介：曹欣然（1985—　），女，山东烟台，农艺师，硕士，从事农作物有害生物控制技术推广工作。xinran1001@163.com。

* **通讯作者**：田明英，推广研究员，从事农作物有害生物控制技术推广工作。E-mail：yttmy5818@sina.com。

海南辣椒病毒种类调查及分子鉴定

余乃通，梁洁，章绍延，王健华，刘志昕*
（中国热带农业科学院热带生物技术研究所，海口 571101）

辣椒是海南冬季瓜菜主要产业之一，也是南菜北运蔬菜的主要品种之一。病毒病的发生制约海南辣椒种植业的发展。为此，我们对海南辣椒病毒病的发生情况进行调查。通过RT-PCR方法对采集的114份疑似病毒辣椒样品进行了检测，共鉴定出5种主要病毒，包括黄瓜花叶病毒（*Cucumber mosaic virus*，CMV）、辣椒轻斑驳病毒（*Pepper mild mottle virus*，PMMoV）、辣椒脉斑驳病毒（*Chilli veinal mottle virus*，ChiVMV），辣椒环斑病毒（*Chilli ringspot virus*，ChiRSV）以及甜椒脉斑驳病毒（*Pepper veinal mottle virus*，PVMV），后3种病毒均为*Potyvirus*属病毒。分析结果表明：海南省辣椒病毒复合侵染现象严重，约有50%的病毒感染辣椒样品是由2种及2种以上病毒共同侵染。

基金项目：1. 海南省重点科技计划应用研究及产业化项目，项目编号：ZDXM20130046；
2. 海南省自然科学基金项目，项目编号：313076。

作者简介：余乃通，男，助研，植物病毒分子生物学，E-mail：yunaitong@163.com。
* **通讯作者**：刘志昕，男，研究员，植物病毒分子生物学，Email：liuzhixin@itbb.org.cn。